Damage Analysis for Composite Materials: Methods, Testing and Evaluation

Damage Analysis for Composite Materials: Methods, Testing and Evaluation

Editors

Luís Miguel Pereira Durão
Nuno Calçada Loureiro

MDPI

Basel • Beijing • Wuhan • Barcelona • Belgrade • Novi Sad • Cluj • Manchester

Editors
Luís Miguel Pereira Durão
Mechanical Engineering
ISEP—Polytechnic of Porto
Porto
Portugal

Nuno Calçada Loureiro
Industrial Production
Engineering
ISVOUGA, Instituto Superior
de entre Douro e Vouga
Santa Maria da Feira
Portugal

Editorial Office
MDPI
St. Alban-Anlage 66
4052 Basel, Switzerland

This is a reprint of articles from the Special Issue published online in the open access journal *Materials* (ISSN 1996-1944) (available at: https://www.mdpi.com/journal/materials/special_issues/P8OSKFE0GW).

For citation purposes, cite each article independently as indicated on the article page online and as indicated below:

Lastname, A.A.; Lastname, B.B. Article Title. *Journal Name* **Year**, *Volume Number*, Page Range.

ISBN 978-3-7258-1311-7 (Hbk)
ISBN 978-3-7258-1312-4 (PDF)
doi.org/10.3390/books978-3-7258-1312-4

Contents

About the Editors

Luís Miguel Pereira Durão

Luís Miguel Pereira Durão has a PhD in Mechanical Engineering. Presently, he is a Coordinator Professor at ISEP - Dept. of Mechanical Engineering, Polytechnic of Porto and develops research activities at INEGI and CIDEM, Porto, Portugal.

Nuno Calçada Loureiro

Prof. Dr. Nuno Calçada Loureiro has a PhD in Leaders for Technical Industries. Presently, he is a Coordinator Professor at ISVOUGA, Dept. of Industrial Production Engineering, Santa Maria da Feira, Portugal and is a researcher at INEGI, Porto, Portugal.

materials

MDPI

Editorial

Special Issue: Damage Analysis for Composite Materials: Methods, Testing and Evaluation

Luis M. P. Durão [1,2,*] and Nuno C. Loureiro [1,3]

1 INEGI Inst. de Ciência e Inovação em Eng. Mecânica e Eng. Industrial, R. Dr. Roberto Frias, 400, 4200-465 Porto, Portugal; n.loureiro@doc.isvouga.pt
2 ISEP, Instituto Politécnico do Porto, R. Dr. António Bernardino de Almeida, 4249-015 Porto, Portugal
3 ISVOUGA, Instituto Superior de Entre Douro e Vouga, R. António de Castro Corte Real, Apart. 132, 4520-181 Santa Maria da Feira, Portugal
* Correspondence: lmd@isep.ipp.pt

After the Stone, the Bronze, and the Iron Ages, material history is now in the Composite Age. Composite materials consist of two or more constituent materials with significantly different physical or chemical properties, which, when combined, produce a material with enhanced properties such as high strength, stiffness, and low weight. Thanks to these unique characteristics, composites are widely used in the production of parts for a large number of final products. In the production process of composite parts, circular hole drilling may be necessary to allow their assembly and incorporation into complex sets, using screws, bolts, pins, rivets, or snap springs. These operations are currently carried out by machining, as drilling is widely used. In the drilling process, delamination is the most frequent damage and can reduce the load-bearing capacity of the parts. The optimization of drilling parameters, the development of specific tool geometries, and the consequences in terms of damage caused by machining have been extensively studied. However, there are still some open issues addressing damage analysis and the assessment of the structural integrity outcomes on load-bearing capacity, fatigue behaviour, or other loading conditions, that deserve research and discussion on parts performance. On the whole, damage analysis for composite materials can be considered a multidisciplinary field that combines materials science, mechanics, numerical analysis, and experimental techniques to ensure the safe and efficient utilisation of composite structures in various applications.

This Special Issue on Damage Analysis for Composite Materials aims to bring to the reader some of the most recent advanced contributions on the study of the outcomes of damage in composite parts, including modelling techniques, failure modes, damage initiation and propagation, durability and fatigue analysis, damage tolerance and repair, non-destructive evaluation, post-failure analysis, and environmental concerns, representing the diverse potential approaches to the study of composite materials. As Guest Editors, it was an honour to be able to bring out all of these contributions in the form of original research articles, showing the ongoing work to broaden the knowledge on this subject.

We cordially invite you to read the papers as they are an interesting set of contributions for the discussion on the subject of damage analysis for composite materials. In [1], the focus was on the interfacial shear strength of carbon nanotubes using finite element methods together with molecular dynamics (MD) and molecular mechanics (MM) to study the interfacial shear strength (ISS) of a structure formed by a pristine single-walled CNT (SWCNT) inserted in a tobermorite crystal. An experimental study on the fracture behaviour of repaired honeycomb/carbon–epoxy sandwich panels under edgewise compression and three-point bending loading was analysed in [2] and complemented by a three-dimensional finite element analysis incorporating a mixed-mode I + II + III cohesive zone model for estimating the fracture behaviour of sandwich panel repairs. In [3], a newly developed failure theory was applied to predict the loads that would result in the fracture of test specimens

Citation: Durão, L.M.P.; Loureiro, N.C. Special Issue: Damage Analysis for Composite Materials: Methods, Testing and Evaluation. *Materials* **2024**, *17*, 2314. https://doi.org/10.3390/ma17102314

Received: 30 April 2024
Accepted: 10 May 2024
Published: 14 May 2024

with holes manufactured from polylactic acid (PLA), subjected to uniaxial loading, and specimens made of polycarbonate (PC), subjected to combined loading. The assessment of the damage caused by drilling holes using enhanced radiography and the consequent outcomes on the fatigue resistance by the cyclic loading of carbon/epoxy plates with the purpose of establishing an association between the damaged region and the material's fatigue is presented in [4]. The fatigue failure of carbon/epoxy composites is the focus in [5], proposing an entropy-based failure criterion to investigate the fatigue lifetime of unidirectional CFRPs subjected to multiple-amplitude cyclic loadings. Due to the heterogeneity of CFRPs, a micro-finite element model that considers matrix resin and fibres independently is developed. Non-destructive techniques are used in [6] to investigate the damage mechanism and energy absorption characteristics of E-glass laminates and sandwich structures with GFRP face sheets with PVC cores under quasi-static indentation with conical, square, and hemispherical indenters with acoustic emission. Afterwards, a postmortem damage assessment was performed with X-ray micro-computed tomography and scanning electron microscopy. In [7], the focus is on showing that Projection-based Digital Volume Correlation (P-DVC) allows 4D (i.e., space and time) full-field measurements that should be carried out over entire loading histories, enabling the quantification of damage detection and growth over the entire loading history up to failure in a polymer matrix composite. Environmental concerns are presented in [8] on a study of the mechanical outcomes due to machining by flexural testing, showing that the use of alternative formulations with micro-inclusions from recovered waste can contribute both to the reduction in the mechanical degradation of composites and to the environment by avoiding the increase in landfill waste. In [9], the Kolmogorov–Sinai metric entropy is used to determine the transition from the elastic to the viscoelastic state in GFRP composite materials during a static tensile test. Additionally, the acoustic emission method is used, showing that it is possible to determine additional parameters affecting the strength of the structure for any composite materials. Finally, in [10], experimental and finite-element simulation are used to study the stability and failure phenomena of thin-walled constructions subjected to axial compression, featuring a central cut-out, and constructed from composite materials. The comparison underscored a significant concordance between the simulation predictions and the empirical findings.

Since November 2022, this Special Issue has attracted the interest of respected researchers from all over the world. The success of this Issue is evidenced by the publication of 10 papers that underwent a rigorous review process conducted by experts in the field. The Guest Editors would like to congratulate all of the authors of the published works and also thank the reviewers for their time and very valuable comments and suggestions that raised the rank and substantive value of this Special Issue. The success of this Special Issue would not be possible without the constant contact and kind support of the Section Managing Editor, Ms. Serena Shi, who should be gratefully thanked for her dedication and commitment. The Guest Editors also thank the Editors-in-Chief of *Materials* for this opportunity to collaborate on the journal and congratulate them on their stewardship of a globally respected journal. The diligence, creativity, and friendly and dynamic cooperation of all those mentioned above contributed to the success of this Special Issue.

Author Contributions: Conceptualization, L.M.P.D. and N.C.L.; writing—original draft preparation, L.M.P.D.; writing—review and editing, N.C.L. All authors have read and agreed to the published version of the manuscript.

Funding: This research received no external funding.

Acknowledgments: The Guest Editors wish to acknowledge all of the authors for their upcoming vital contributions to this Special Issue, and to the editorial staff of *Materials* for their outstanding support.

Conflicts of Interest: The authors declare that they have no known competing financial interests or personal relationships that could have appeared to influence the work reported in this paper.

References

1. Talayero, C.; Lado-Touriño, I.; Aït-Salem, O.; Ramos, I.S.; Páez-Pavón, A.; Merodio-Perea, R.G. Interfacial Shear Strength of Single-Walled Carbon Nanotubes-Cement Composites from Molecular Dynamics and Finite Element Studies. *Materials* **2023**, *16*, 1992. [CrossRef] [PubMed]
2. Rocha, R.J.B.; de Moura, M.F.S.F.; Moreira, R.D.F. Edgewise Compression and Three-Point Bending Analyses of Repaired Composite Sandwich Panels. *Materials* **2023**, *16*, 4249. [CrossRef] [PubMed]
3. Schmeier, G.E.C.; Tröger, C.; Kwon, Y.W.; Sachau, D. Predicting Failure of Additively Manufactured Specimens with Holes. *Materials* **2023**, *16*, 2293. [CrossRef] [PubMed]
4. Durão, L.M.P.; Matos, J.E.; Loureiro, N.C.; Esteves, J.L.; Fernandes, S.C.F. Damage Propagation by Cyclic Loading in Drilled Carbon/Epoxy Plates. *Materials* **2023**, *16*, 2688. [CrossRef] [PubMed]
5. Deng, H.; Toda, K.; Sato, M.; Koyanagi, J. Micro-Scale Numerical Simulation of Fatigue Failure for CFRP Subjected to Multiple-Amplitude Cyclic Loadings Based on Entropy Damage Criterion. *Materials* **2023**, *16*, 6120. [CrossRef] [PubMed]
6. Osa-uwagboe, N.; Udu, A.G.; Silberschmidt, V.V.; Baxevanakis, K.P.; Demirci, E. Damage Assessment of Glass-Fibre-Reinforced Plastic Structures under Quasi-Static Indentation with Acoustic Emission. *Materials* **2023**, *16*, 5036. [CrossRef] [PubMed]
7. Mandić, A.; Kosin, V.; Jailin, C.; Tomičević, Z.; Smaniotto, B.; Hild, F. Damage Detection in a Polymer Matrix Composite from 4D Displacement Field Measurements. *Materials* **2023**, *16*, 6300. [CrossRef] [PubMed]
8. Durão, L.M.P.; Matos, J.E.; Alves, J.; Filho, S.M.R.; Panzera, T.H.; Scarpa, F. Experimental Study of Drilling Damage Outcomes in Hybrid Composites with Waste Micro-Inclusions. *Materials* **2023**, *16*, 7325. [CrossRef] [PubMed]
9. Panasiuk, K.; Dudzik, K.; Hajdukiewicz, G.; Abramczyk, N. Acoustic Emission and K-S Metric Entropy as Methods to Analyze the Influence of Gamma-Aluminum Oxide Nanopowder on the Destruction Process of GFRP Composite Materials. *Materials* **2023**, *16*, 7334. [CrossRef] [PubMed]
10. Falkowicz, K. Stability and Failure of Thin-Walled Composite Plate Elements with Asymmetric Configurations. *Materials* **2024**, *17*, 1943. [CrossRef] [PubMed]

materials

MDPI

Article

Interfacial Shear Strength of Single-Walled Carbon Nanotubes-Cement Composites from Molecular Dynamics and Finite Element Studies

Carlos Talayero [1,*], Isabel Lado-Touriño [1], Omar Aït-Salem [2], Ismael Sánchez Ramos [2], Alicia Páez-Pavón [1] and Rosario G. Merodio-Perea [1]

[1] Engineering Department, School of Architecture, Engineering and Design, Universidad Europea de Madrid, 28670 Villaviciosa de Odón, Spain
[2] Hexagon HMI, 28050 Madrid, Spain
* Correspondence: carlosalberto.talayero@universidadeuropea.es

Abstract: Carbon nanotubes (CNTs) are nanometer-sized structures that can be used to reinforce cement matrices. The extent to which the mechanical properties are improved depends on the interfacial characteristics of the resulting materials, that is, on the interactions established between the CNTs and the cement. The experimental characterization of these interfaces is still impeded by technical limitations. The use of simulation methods has a great potential to give information about systems lacking experimental information. In this work, molecular dynamics (MD) and molecular mechanics (MM) were used in conjunction with finite element simulations to study the interfacial shear strength (ISS) of a structure formed by a pristine single-walled CNT (SWCNT) inserted in a tobermorite crystal. The results show that, for a constant SWCNT length, ISS values increase when the SWCNT radius increases, while for a constant SWCNT radius, shorter lengths enhance ISS values.

Keywords: carbon nanotubes; interfacial shear strength; pull-out test; finite element method; molecular model

Citation: Talayero, C.; Lado-Touriño, I.; Aït-Salem, O.; Ramos, I.S.; Páez-Pavón, A.; G. Merodio-Perea, R. Interfacial Shear Strength of Single-Walled Carbon Nanotubes-Cement Composites from Molecular Dynamics and Finite Element Studies. *Materials* **2023**, *16*, 1992. https://doi.org/10.3390/ma16051992

Academic Editor: Alexander N. Obraztsov

Received: 6 February 2023
Revised: 23 February 2023
Accepted: 26 February 2023
Published: 28 February 2023

1. Introduction

Cement and cement composites are the most used materials in the building sector due to their high compressive strength, low prices and mature production technology. However, cement also has low tensile strength, good permeability to several substances and easily cracks. To improve these weaknesses, different types of reinforcements are added to cement. Carbon nanotubes (CNTs) are one of the most promising candidates due to their excellent mechanical properties. The number of research studies on cement-CNTs composites has steadily grown in recent years [1–5]. CNTs have elastic moduli greater than 1 TPa and tensile strengths as high as 63 GPa [6,7]. When CNTs are homogeneously dispersed within the cement matrix, density [8], permeability [9], mechanical properties [10] and durability [11] of the composites are improved. The effect of CNTs is particularly noticeable on mechanical properties, as they can delay cracks propagation.

The CNTs—cement interface is a key factor in mechanical properties as this is typically a weak area and determines the material fracture behavior [12]. Diverse simulation methods, such as coarse-grained and all-atoms molecular dynamics [13], ref. [14] show that when CNTs are incorporated into a cement matrix, the way of fracture is changed and mechanical properties are clearly enhanced.

Interfacial bond strength can be determined from Interfacial Shear Strength (ISS) measurements, such as fragmentation, pull-out, microdroplet, push-out and push-in tests [15–20] and most experimental studies have focused on composites made of CNTs and polymeric matrices due to the difficulties in studying CNT-cement composites. The lack of experimental information can be supplemented by the use of simulation methods such as molecular mechanics (MM) and molecular dynamics (MD). Both techniques provide atomic information about the interface, which cannot be obtained from experiments. ISS can be calculated using a pull-out model [21]. Pulling out a CNT from a matrix provide information about the interface strength. There are numerous molecular modelling studies of ISS of CNTs-reinforced polymer composites [22–26], while results on carbon-based reinforcements are scarce and most of them concern graphene [27–30].

The Finite Element Method (FEM) is other type of study that can be used to calculate mechanical properties of composite materials. It can be based on different formulations, such as beam models [31], representative volume element FEM analysis [3,32] or nonlinear analysis for dynamic events [33] or when crack analysis is needed [34].

In this work, SWCNT pull-out simulations from a tobermorite crystal were performed to calculate the ISS of several nanocomposites. MD, MM were applied to obtain interfacial energies, which were subsequently used as input in a FE calculation. Pristine SWCNTs with different radii (2.70 to 4.74 Å) and lengths varying from 9.84 Å to 19.68 Å, were inserted into a tobermorite were explained as a function of SWCNT diameter and length and related to non-bonded interactions between the SWCNT and cement. Our work aims to get more insight into the effect that SWCNT geometry plays on ISS, which is a key factor for enhancing mechanical properties of these materials. We think that the synergy achieved by using both atomistic and FE methods is a powerful way of attaining our goals.

2. Models and Methods

In this section, the modelling methods are described.

2.1. Atomistic Model Systems

The periodic simulation cell used for the pull-out atomistic calculations had dimensions of $29.54 \times 67.46 \times 500$ Å^3. The cell in the pull-out direction was large enough and included a vacuum layer (Figure 1).

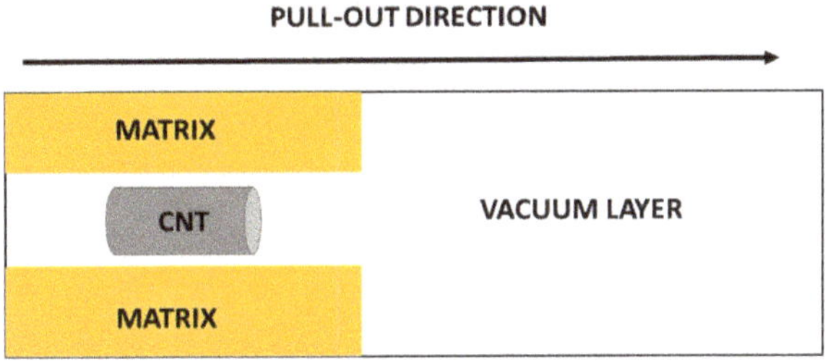

Figure 1. Cell used for the atomistic pull-out simulations.

The atomic models built to represent the cement matrix and the composite material are shown in Figure 2. Despite the fact that cementitious systems are not fully crystalline [35], tobermorite 11 Å was employed to model cement by many different authors [36–38]. As we are mainly interested in studying interfacial interactions, crystalline tobermoritecan be a good approximation to this interface.

To build the model of the composite material, a SWCNT was inserted into the tobermorite crystal (Figure 2). Different pristine SWCNTs ((CNT (4, 4) = 2.70 Å, CNT (5, 5) = 3.39 Å, CNT (6, 6) = 4.07 Å and CNT (7, 7) = 4.74 Å) with lengths of 9.64 Å, 19.68 Å and 24.60 were used to study the influence of diameter and length on ISS.

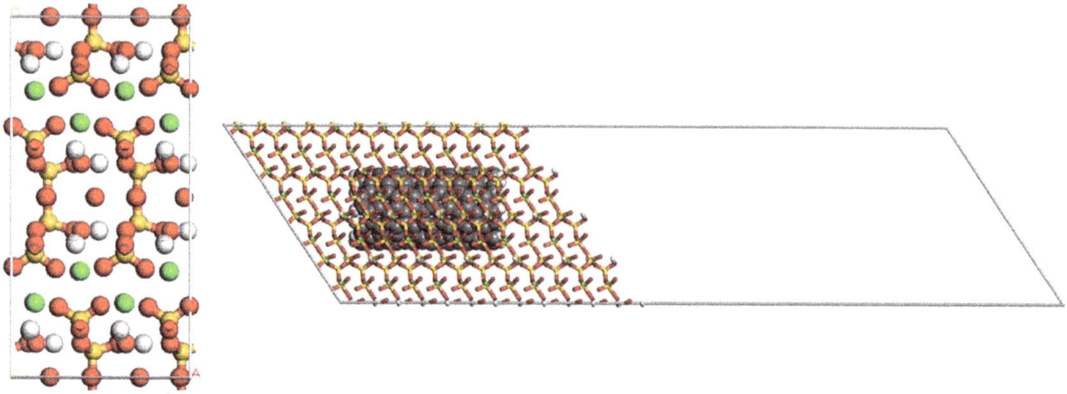

Figure 2. Models used in the simulation; cement: tobermorite crystal (left) and composite: tobermorite/SWCNT.

2.2. Molecular Mechanics and Molecular Dynamics

The first step in creating a good model for the pull-out process was to relax the simulation lattice by means of a 100 ps simulation in the NPT ensemble at 298 K and atmospheric pressure. The temperature and pressure were controlled by a Nose-Hoover thermostat [39] and a Berendsen barostat [40] respectively. A time step of 1 fs was used during the simulation. Three independent simulations were carried out to obtain average values. After relaxing the structures, they were used to simulate the pull-out process. The size of the cell in x direction was increased and a vacuum layer was included. Before the pull-out study, these new models were subjected to a process of optimization by molecular mechanics (MM). Then, pulling-out was done by displacing the SWCNT along the x axis with increments of 10 Å. Initial and final structures before and after the SWCNT pull-out process are shown in Figure 3. After each pull-out step, the energy of the system was calculated by MM. Materials Studio 7.0 software [41] was used to perform the simulations.

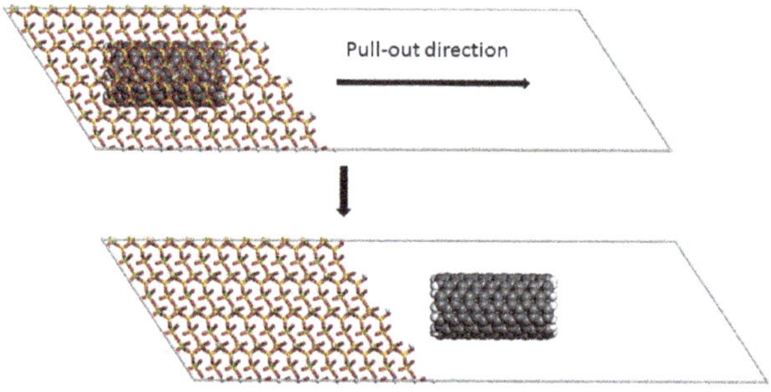

Figure 3. Initial and final structures before (top) and after (bottom) the SWCNT pull-out process.

The condensed-phase optimized molecular potential for atomistic simulation studies forcefield COMPASSII [42] was employed to calculate the potential energy of all systems. This forcefield has been successfully applied in numerous MD studies of systems containing CNTs and cement derived materials [28,43–46]. Coulomb interactions were calculated by the Ewald summation method while an atom-based cutoff method was applied for van der Waal interactions. The cutoff distance for both interactions was 12.5 Å.

The interfacial energy ($E_{interfacial}$) was calculated as the energy difference between the fully embedded configuration of the SWCNT and the complete pulled-out configuration (Figure 3). This energy was then used to calculate the ISS by FEM.

2.3. Finite Element Method Model

A Finite Element Model based on Axisymmetric conditions has been built in MSC.Marc software by HEXAGON [47] to represent the composite material. Different FE models have been defined to cover the different combinations of radius and lengths. Same loads, boundary conditions, physical and material properties and analysis methodology are used for each geometrical combination.

2.3.1. Loads & Boundary Conditions

Loads and Boundary Conditions were defined using contact bodies and contact interactions. Five contact bodies were defined (Figure 4): Armchair (a flexible body corresponding to Pristine armchair SWCNTs material), tobermorite (a flexible body corresponding totobermoritematrix material, Base (a rigid body defined by a curve), Tirador (a rigid body defined by a curve) and Sym_Axy (a rigid body defined by a curve). The contact interactions defined between contact bodies represented the load and boundary conditions where "Tirador–Armchair" was a glue contact interaction. This interaction defined the load as Tirador rigid body and it was controlled by its position. Tirador body was moved 1.1×10^{-9} mm in X-Axis, so it pulled Armchair body due to glue contact interaction. The "Sym_Axy–Armchair" contact was defined as a touching contact interaction. This interaction represents the axisymmetric boundary condition. The "Base–Armchair" pair was defined as a glue contact interaction. This interaction represented a fixed boundary condition. Finally, the "Base–tobermorite" interaction was defined as a glue contact interaction. This interaction represented a fixed boundary condition.

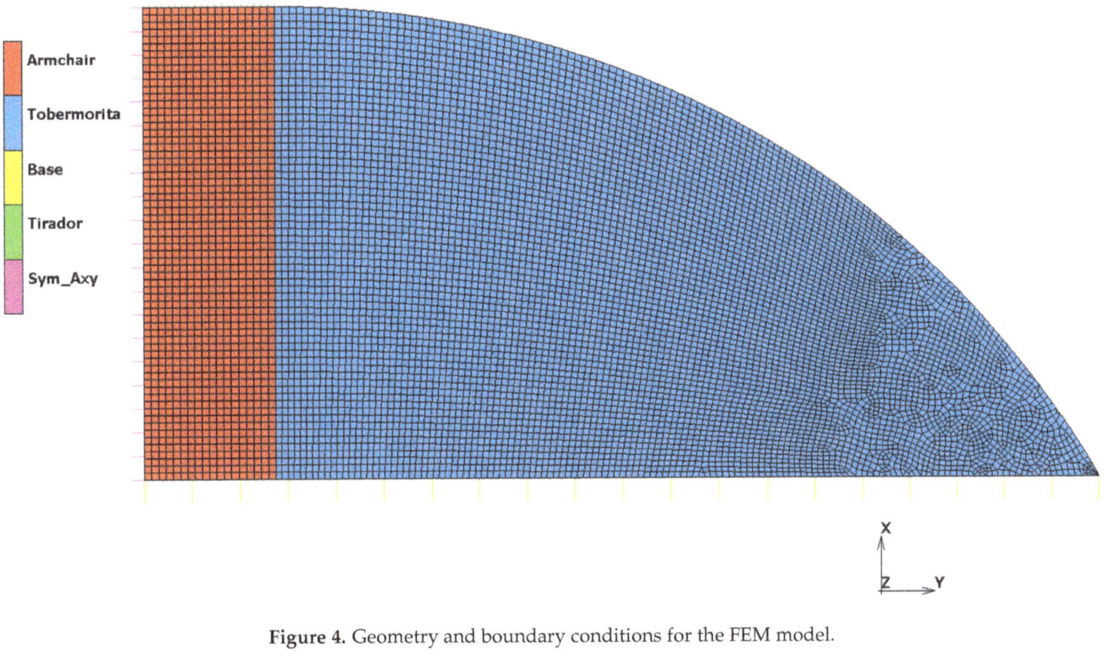

Figure 4. Geometry and boundary conditions for the FEM model.

Furthermore, a delamination condition was defined in the interface between Armchair and tobermorite elements. This delamination condition automatically created interface cohesive elements (CZM) between both components when stresses were higher than the stablished (10^{-9} MPa). Very low values (numerically zero) were set to allow the software Marc always automatically split the mesh between Armchair and tobermorite elements, and introduce interface cohesive elements.

2.3.2. Physical Properties

Axisymmetric properties were defined for all elements of the model, for both Armchair and tobermorite components. Marc Element types 10 (Arbitrary Quadrilateral Axisymmetric Ring) and 2 (Axisymmetric Triangular Ring) were used [48].

2.3.3. Material Properties

Armchair and tobermorite were defined using standard isotropic elastic material model. The Young's Modulus (E) of the Armchair is set to 98,413 MPa and the Poisson Ratio (υ) to 0.13. The Young's Modulus (E) of the tobermorite is set to 85,000 MPa and the Poisson Ratio (υ) to 0.3.

Interface Cohesive elements (CZM) were defined using Exponential formulation with the following properties derived from the where the values obtained with MM and MD methods in order to get a coherent traction vs opening displacement curve: Cohesive Energy (G_c) of 9.64×10^{-9} mJ and Critical Opening Displacement (v_c) of 7.81×10^{-11} mm.

The effective traction (t) was introduced as a function of the effective opening displacement (v) and was characterized by an initial reversible response followed by an irreversible response as soon as a critical effective opening displacement (v_c) has been reached. The irreversible part was characterized by increasing damage ranging from 0 (onset of delamination) to 1 (full delamination). The exponential formulation has the following Equation (1) [48].

$$t = G_c \frac{v}{v_c^2} e^{\frac{-v}{v_c}} \quad (1)$$

in which G_c is the energy release rate (cohesive energy).

2.3.4. Analysis Methodology

An Implicit Non-Linear analysis using MSC.Marc is performed using a constant load step (1% of total load is applied in each load step), a convergence method based on residual forces criterion with automatic switching to displacement criterion was used, a relative convergence force tolerance of 10% was defined, large strain formulation was set and "Node-to-Segment" contact algorithm was selected.

3. Results and Discussion

3.1. Interfacial Energy

The variation in total potential energy with pull-out displacement for the model containing a SWCNT(6, 6) is shown in Figure 5. As the SWCNT was pulled-out from the matrix, the total potential energy increased and reached a constant value beyond a certain pull-out distance. This indicated that the SWCNT and the matrix no longer interacted. The energy of the fully embedded configuration was lower than the energy of the complete pulled-out configuration. Many other authors found similar results [23,26–28]. The interfacial energy is the energy difference between the fully embedded configuration of the SWCNT and the complete pulled-out configuration (Figure 5).

The interfacial energy values are presented in Figure 6 for all systems. The standard deviation of each value is also shown in this figure. In general, interfacial energy increases when SWCNT diameter and length increase. This is probably due to the greater number of atoms in longer and bigger SWCNTs, which increases the contact points between the SWCNT and the tobermorite and, hence, the interfacial energy.

3.2. Interfacial Shear Strength

For each FE model, Contact Force in X-Axis of contact body Base can be plotted with respect to position in X-Axis of contact body Tirador. The maximum value of this force can be considered as the limit value from which delamination begins to occur. This force will be used to calculate ISS dividing it by contact area between Armchair and tobermorite contact bodies (Figure 7).

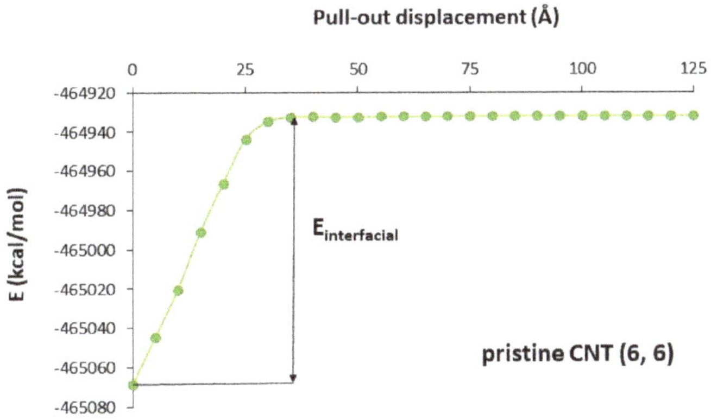

Figure 5. Total potential energy versus pull−out displacement.

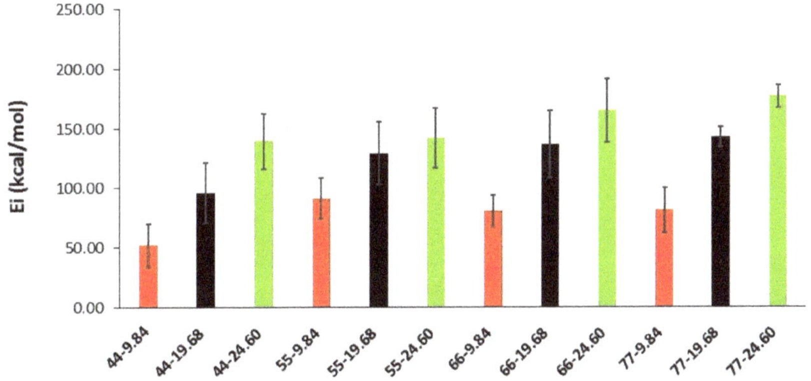

Figure 6. Interfacial energy as a function of SWCNT diameter and length.

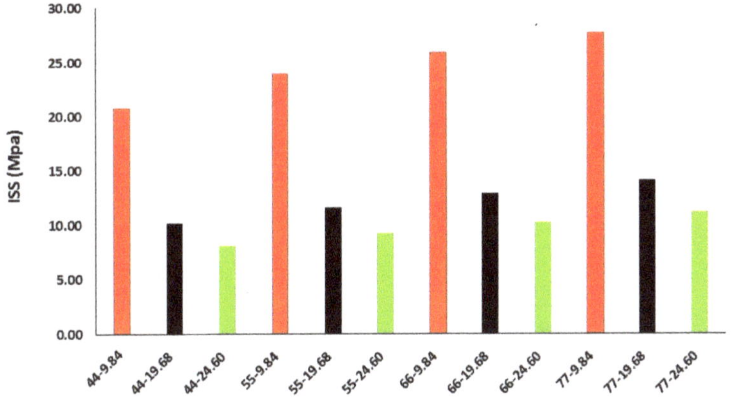

Figure 7. ISS as a function of SWCNT diameter and length.

ISS values are improved when SWCNT radius increases, and SWCNT length shortens. This is given by a similar extraction force (Table 1) for each diameter that leads to lower stress values when the contact area increases (length). This trend was also found by Li et al. [24] for pristine SWCNTs/polyethylene matrices. They attributed the positive correlation between SWCNT dimensions and ISS values to the increase of interfacial atoms.

Table 1. Contact force and ISS of the different systems.

Radius (Å)	Length (Å)	Maximum Contact Force ($N \times 10^{-11}$)	Contact Area ($mm^2 \times 10^{-12}$)	ISS (MPa)
2.71	9.84	3.48	1.68	20.740
	19.68	3.40	3.35	10.158
	24.6	3.39	4.19	8.091
3.39	9.84	5.02	2.10	23.932
	19.68	4.87	4.19	11.620
	24.6	4.83	5.24	9.210
4.07	9.84	6.51	2.52	25.859
	19.68	6.49	5.03	12.896
	24.6	6.44	6.29	10.237
4.74	9.84	8.09	2.93	27.602
	19.68	8.27	5.86	14.115
	24.6	8.20	7.33	11.194

It can be seen from Table 1 that, depending upon the SWCNT length and diameter, calculated ISS values ranged from 8.09 to 27.70 MPa. The effect of CNTs length on mechanical properties is controversial, as its efficiency as reinforcing agents depends on several factors such as defects on the CNT surface, concentration or dispersion in the cement matrix, although, in general, the addition of CNTs to cement leads to high improvements in mechanical properties. Some authors found that long CNTs performed better, and, to improve mechanical properties, highest amounts of short CNTs were needed [49,50].

Other authors [51] found that composites reinforced with the long CNTs showed worse properties and they attributed the efficiency of the shortest CNTs to a better filling of the nanopores within the cement matrix.

In a recent work, it was observed that the CNT length had a minimal effect on the properties of cement [52], as it is obtained in our simulations. For a given radius, the contact force can consider as constant, resulting in higher ISS for smaller areas (longer CNTs).

Thus, to validate our approach, a comparison of our results with experimental data of single SWCNT pull-out from cement matrices would be extremely useful. To the best of our knowledge, none is currently available as the experimental characterization of these interfaces still faces numerous technical problems.

It is noticeable that, for a given radius of the SWCNT, the energy rises with the length (increase of interfacial atoms) but the ISS decreases (the contact area is higher, and the ISS is obtained by dividing the force by the area).

3.3. *Damage Level*

If a more detailed post-processing is done in the FE models, a Damage Level can be studied. Damage value indicates the amount of irreversible cohesive energy that has been lost. A value of 1.0 indicates complete loss of cohesive energy and an area where delamination will begin to occur. Damage results can be defined as follows [48]:

$$D = \frac{G - G_{elastic}}{G_c - G_{elastic}}; 0 \leq D \leq 1; v_e \geq v_c \quad (2)$$

For instance, in the case of Radius 2.71 Å and Length 9.84 Å, delamination starts when cohesive elements begin to be damaged at 24% of total load applied (Displacement × Tirador: 2.64×10^{-10} mm).

In Figure 8a it can be observed that crack initiates in the interface between Armchair and tobermorite when Cohesive Interface Elements begin to have irreversible behavior (2.238 % of Damage Level). It must be noticed that deformed shape is not in real scale but scaled with a factor in order to be able to see Damage Level in the Cohesive Interface Elements.

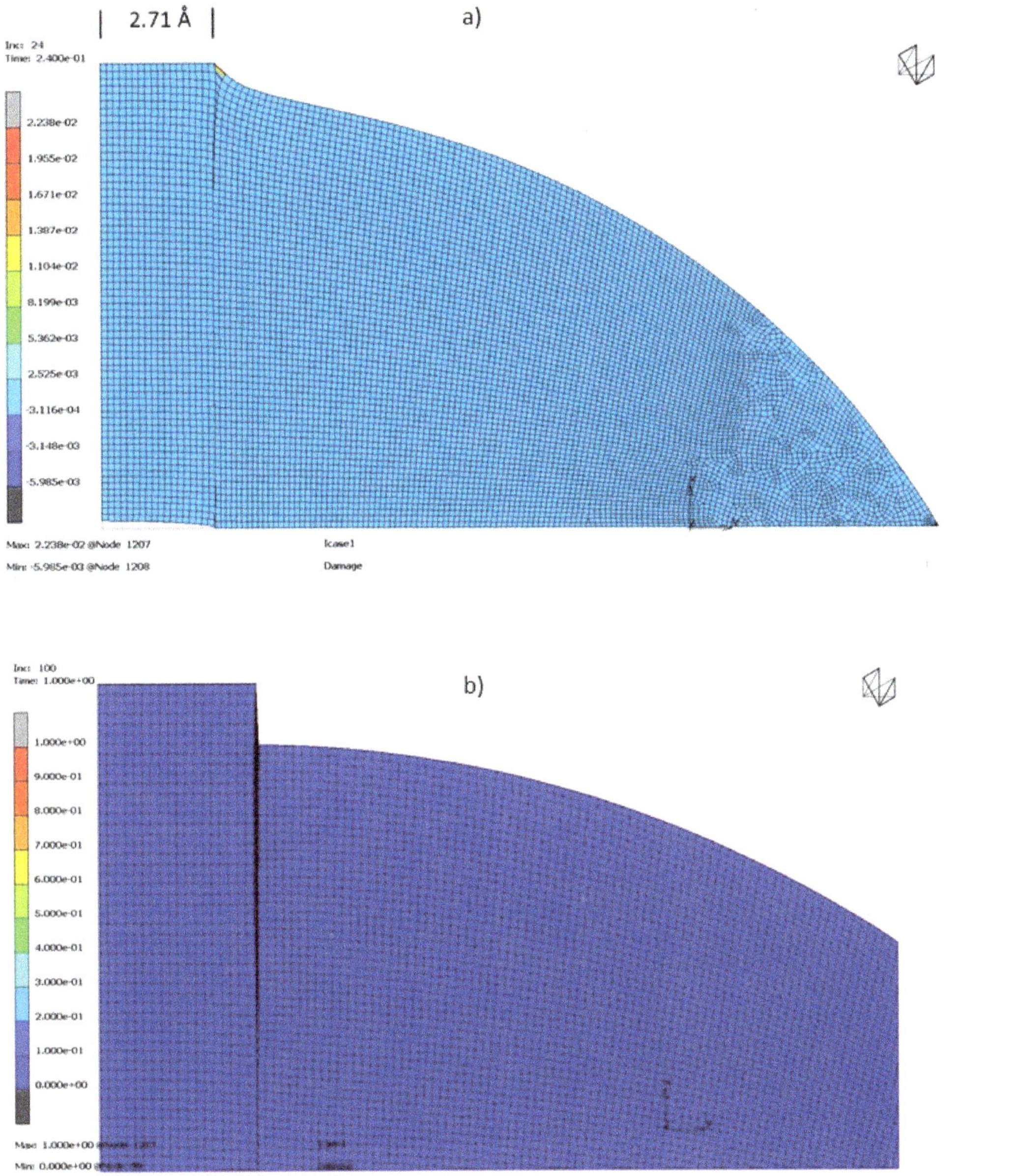

Figure 8. Damage level for for Radius 2.71 Å and Length 9.84 Å (**a**) Damage level when delamination starts. (**b**) Damage level when damage reaches 100%.

At the end of the analysis (Figure 8b)) (Displacement × Tirador: 1.1×10^{-9} mm), it can be observed how the delamination has grown through the interface between Armchair and tobermorite and how Damage Level of Cohesive Interface Elements have also grown until 100% of Damage Level (deformed shape is also scaled with a magnification factor).

The Damage level analysis permits to confirm that the pull-out test results are affected by the cohesion energy given by molecular models. On the other hand, the results of the test are limited by the weakest element, obtaining ISS values coherent with the cement-based materials limits [53].

4. Conclusions

Pristine SWCNTs with different geometrical characteristics were pulled-out from a tobermorite matrix to study the interfacial characteristics of the resulting composites by MM and MD simulations. Only non-bonded interactions between the SWCNT and the cement matrix were considered. The interfacial energy obtained from these atomistic simulations were subsequently used as input in FEM calculations to determine ISS values. ISS values increased with larger SWCNT radius, which could be attributed to a more extensive contact surface between the SWCNT and the tobermorite. However, ISS showed an opposite trend, being larger for shorter CNTs. This is due to the increased area for longer CNTs for an almost constant force for a given radius. The weakest zone of the material seems to be the tobermorite matrix. Delamination and crack growth is mainly observed at the tobermorite side of the interface. Thus, the introduction of a SWCNT of varying geometry modifies the mechanical behavior of the matrix as confirmed by the Damage Level analysis test. The pull-out results are affected by the cohesion energy, which, in turn, depends on the CNT geometry. We think the joint use of atomistic and FEM methods seems to be a good approach to describe the interfacial properties of this kind of material. It would be very interesting to be able to compare the results of this simulation with experiments, but, to our knowledge, none is currently available.

Author Contributions: C.T., O.A.-S. and I.S.R. worked on the FEM simulations and results interpretation. I.L.-T., A.P.-P. and R.G.M.-P. worked on the Molecular models. I.L.-T. and C.T. prepared the manuscript. All authors have read and agreed to the published version of the manuscript.

Funding: The APC was funded by Universidad Europea de Madrid and MSC. Software Simulating Reality, S.A.

Institutional Review Board Statement: Not applicable.

Informed Consent Statement: Not applicable.

Data Availability Statement: All data generated and analyzed during this study are generated by commercial software and the conditions and properties are described in the article. For that reason, they are not available as supplementary information files but are available from the corresponding author on reasonable request.

Conflicts of Interest: The authors declare no conflict of interest.

References

1. Reales, O.A.M.; Filho, R.D.T. A review on the chemical, mechanical and microstructural characterization of carbon nanotubes-cement based composites. *Constr. Build. Mater.* **2017**, *154*, 697–710. [CrossRef]
2. Shi, T.; Li, Z.; Guo, J.; Gong, H.; Gu, C. Research progress on CNTs/CNFs-modified cement-based composites—A review. *Constr. Build. Mater.* **2019**, *202*, 290–307. [CrossRef]
3. Rocha, V.V.; Ludvig, P. Influence of Carbon Nanotubes on the Mechanical Behavior and Porosity of Cement Pastes Prepared by A Dispersion on Cement Particles in Isopropanol Suspension. *Materials* **2020**, *13*, 3164. [CrossRef] [PubMed]
4. Makul, N. Advanced smart concrete—A review of current progress, benefits and challenges. *J. Clean. Prod.* **2020**, *274*, 122899. [CrossRef]
5. Metaxa, Z.; Tolkou, A.; Efstathiou, S.; Rahdar, A.; Favvas, E.; Mitropoulos, A.; Kyzas, G. Nanomaterials in Cementitious Composites: An Update. *Molecules* **2021**, *26*, 1430. [CrossRef] [PubMed]
6. Li, G.Y.; Wang, P.M.; Zhao, X. Mechanical behavior and microstructure of cement composites incorporating surface-treated multi-walled carbon nanotubes. *Carbon N. Y.* **2005**, *43*, 1239–1245. [CrossRef]

7. Li, G.Y.; Wang, P.M.; Zhao, X. Pressure-sensitive properties and microstructure of carbon nanotube reinforced cement composites. *Cem. Concr. Compos.* **2007**, *29*, 377–382. [CrossRef]
8. Gao, F.; Tian, W.; Wang, Z.; Wang, F. Effect of diameter of multi-walled carbon nanotubes on mechanical properties and microstructure of the cement-based materials. *Constr. Build. Mater.* **2020**, *260*, 120452. [CrossRef]
9. Gao, Y.; Jing, H.; Zhou, Z.; Shi, X.; Li, L.; Fu, G. Roles of carbon nanotubes in reinforcing the interfacial transition zone and impermeability of concrete under different water-to-cement ratios. *Constr. Build. Mater.* **2021**, *272*, 121664. [CrossRef]
10. Adhikary, S.K.; Rudžionis, Ž.; Rajapriya, R. The Effect of Carbon Nanotubes on the Flowability, Mechanical, Microstructural and Durability Properties of Cementitious Composite: An Overview. *Sustainability* **2020**, *12*, 8362. [CrossRef]
11. Du, Y.; Gao, P.; Yang, J.; Shi, F.; Shabaz, M. Experimental Analysis of Mechanical Properties and Durability of Cement-Based Composite with Carbon Nanotube. *Adv. Mater. Sci. Eng.* **2021**, *2021*, 8777613. Available online: https://go.gale.com/ps/i.do?p=AONE&sw=w&issn=16878434&v=2.1&it=r&id=GALE%7CA696851757&sid=googleScholar&linkaccess=fulltext (accessed on 6 October 2022). [CrossRef]
12. Yu, Z.; Lau, D. Evaluation on mechanical enhancement and fire resistance of carbon nanotube (CNT) reinforced concrete. *Coupled Syst. Mech.* **2017**, *6*, 335–349. [CrossRef]
13. Qin, R.; Zhou, A.; Yu, Z.; Wang, Q.; Lau, D. Role of carbon nanotube in reinforcing cementitious materials: An experimental and coarse-grained molecular dynamics study. *Cem. Concr. Res.* **2021**, *147*, 106517. [CrossRef]
14. Sindu, B.S.; Sasmal, S. Molecular dynamics simulations for evaluation of surfactant compatibility and mechanical characteristics of carbon nanotubes incorporated cementitious composite. *Constr. Build. Mater.* **2020**, *253*, 119190. [CrossRef]
15. Yilmaz, Y.I. Analyzing Single Fiber Fragmentation Test Data by Using Stress Transfer Model. *J. Compos. Mater.* **2002**, *36*, 537–551. [CrossRef]
16. Zhao, Y.R.; Xing, Y.M.; Lei, Z.K.; Lang, F.C. Interfacial stress transfer behavior in a specially-shaped fiber/matrix pullout test. *Acta Mech. Sin.* **2009**, *26*, 113–119. [CrossRef]
17. Zu, M.; Li, Q.; Zhu, Y.; Dey, M.; Wang, G.; Lu, W.; Deitzel, J.M.; Gillespie, J.W.; Byun, J.-H.; Chou, T.-W. The effective interfacial shear strength of carbon nanotube fibers in an epoxy matrix characterized by a microdroplet test. *Carbon N. Y.* **2012**, *50*, 1271–1279. [CrossRef]
18. Battisti, A.; Ojos, D.E.-D.L.; Ghisleni, R.; Brunner, A.J. Single fiber push-out characterization of interfacial properties of hierarchical CNT-carbon fiber composites prepared by electrophoretic deposition. *Compos. Sci. Technol.* **2014**, *95*, 121–127. [CrossRef]
19. Rodríguez, M.; Molina-Aldareguía, J.; González, C.; Llorca, J. A methodology to measure the interface shear strength by means of the fiber push-in test. *Compos. Sci. Technol.* **2012**, *72*, 1924–1932. [CrossRef]
20. Jean, J.-H.; Yu, C.; Park, C.-G. Bonding Characteristics of Macrosynthetic Fiber in Latex-Modified Fiber-Reinforced Cement Composites as a Function of Carbon Nanotube Content. *Int. J. Polym. Sci.* **2016**, *2016*, 7324975. [CrossRef]
21. Li, Y.; Wang, Q.; Wang, S. A review on enhancement of mechanical and tribological properties of polymer composites reinforced by carbon nanotubes and graphene sheet: Molecular dynamics simulations. *Compos. Part B Eng.* **2019**, *160*, 348–361. [CrossRef]
22. Liao, K.; Li, S. Interfacial characteristics of a carbon nanotube–polystyrene composite system. *Appl. Phys. Lett.* **2001**, *79*, 4225–4227. [CrossRef]
23. Chowdhury, S.; Okabe, T. Computer simulation of carbon nanotube pull-out from polymer by the molecular dynamics method. *Compos. Part A Appl. Sci. Manuf.* **2007**, *38*, 747–754. [CrossRef]
24. Li, Y.; Liu, Y.; Peng, X.; Yan, C.; Liu, S.; Hu, N. Pull-out simulations on interfacial properties of carbon nanotube-reinforced polymer nanocomposites. *Comput. Mater. Sci.* **2011**, *50*, 1854–1860. [CrossRef]
25. Chandra, Y.; Scarpa, F.; Adhikari, S.; Zhang, J.; Flores, E.S.; Peng, H.-X. Pullout strength of graphene and carbon nanotube/epoxy composites. *Compos. Part B Eng.* **2016**, *102*, 1–8. [CrossRef]
26. Chawla, R.; Sharma, S. Molecular dynamics simulation of carbon nanotube pull-out from polyethylene matrix. *Compos. Sci. Technol.* **2017**, *144*, 169–177. [CrossRef]
27. Fan, D.; Lue, L.; Yang, S. Molecular dynamics study of interfacial stress transfer in graphene-oxide cementitious composites. *Comput. Mater. Sci.* **2017**, *139*, 56–64. [CrossRef]
28. Alkhateb, H.; Al-Ostaz, A.; Cheng, A.H.-D.; Li, X. Materials Genome for Graphene-Cement Nanocomposites. *J. Nanomech. Micromech.* **2013**, *3*, 67–77. [CrossRef]
29. Eftekhari, M.; Mohammadi, S. Molecular dynamics simulation of the nonlinear behavior of the CNT-reinforced calcium silicate hydrate (C–S–H) composite. *Compos. Part A Appl. Sci. Manuf.* **2016**, *82*, 78–87. [CrossRef]
30. Kai, M.; Zhang, L.; Liew, K. Graphene and graphene oxide in calcium silicate hydrates: Chemical reactions, mechanical behavior and interfacial sliding. *Carbon N. Y.* **2019**, *146*, 181–193. [CrossRef]
31. Joshi, U.A.; Joshi, P.; Harsha, S.P.; Sharma, S.C. Evaluation of the Mechanical Properties of Carbon Nanotube Based Composites by Finite Element Analysis. *Int. J. Eng. Sci. Technol.* **2010**, *2*, 1098–1107.
32. Talayero, C.; Aït-Salem, O.; Gallego, P.; Páez-Pavón, A.; Merodio-Perea, R.G.; Lado-Touriño, I. Computational Prediction and Experimental Values of Mechanical Properties of Carbon Nanotube Reinforced Cement. *Nanomaterials* **2021**, *11*, 2997. [CrossRef] [PubMed]
33. RAnsari, R.; Hemmatnezhad, M. Nonlinear finite element analysis for vibrations of double-walled carbon nanotubes. *Nonlinear Dyn.* **2011**, *67*, 373–383. [CrossRef]

34. Alhassan, M.; Betoush, N.; Al-Huthaifi, N.; Al Dalou, A. Estimation of the fracture parameters of macro fiber-reinforced concrete based on nonlinear elastic fracture mechanics simulations. *Results Eng.* **2022**, *15*, 100539. [CrossRef]
35. Bauchy, M.; Qomi, M.J.A.; Ulm, F.J.; Pellenq, R.J.M. Order and disorder in calcium–silicate–hydrate. *J. Chem. Phys.* **2014**, *140*, 214503. [CrossRef] [PubMed]
36. Richardson, I.G. Tobermorite/jennite- and tobermorite/calcium hydroxide-based models for the structure of C-S-H: Applicability to hardened pastes of tricalcium silicate, β-dicalcium silicate, Portland cement, and blends of Portland cement with blast-furnace slag, metakaolin, or silica fume. *Cem. Concr. Res.* **2004**, *34*, 1733–1777. [CrossRef]
37. Hou, D.; Lu, Z.; Li, X.; Ma, H.; Li, Z. Reactive molecular dynamics and experimental study of graphene-cement composites: Structure, dynamics and reinforcement mechanisms. *Carbon N. Y.* **2017**, *115*, 188–208. [CrossRef]
38. Allen, A.J.; Thomas, J.J.; Jennings, H.M. Composition and density of nanoscale calcium-silicate-hydrate in cement. *Nat. Mater.* **2007**, *6*, 311–316. [CrossRef] [PubMed]
39. Hoover, W.G. Canonical dynamics: Equilibrium phase-space distributions. *Phys. Rev. A* **1985**, *31*, 1695. [CrossRef]
40. Berendsen, H.J.C.; Postma, J.P.M.; Van Gunsteren, W.F.; Dinola, A.; Haak, J.R. Molecular dynamics with coupling to an external bath. *J. Chem. Phys.* **1998**, *81*, 3684. [CrossRef]
41. BIOVIA Materials Studio—BIOVIA—Dassault Systèmes®. Available online: https://www.3ds.com/products-services/biovia/products/molecular-modeling-simulation/biovia-materials-studio/ (accessed on 6 October 2022).
42. Sun, H. The COMPASS force field: Parameterization and validation for phosphazenes. *Comput. Theor. Polym. Sci.* **1998**, *8*, 229–246. [CrossRef]
43. Nano Mechanical Properties on the Mineralogical Array of Calcium Silicate Hydrates and Calcium Hydroxide through Molecular Dynamics-CSIR-SERC | Request PDF. Available online: https://www.researchgate.net/publication/272702301_Nano_mechanical_properties_on_the_mineralogical_array_of_calcium_silicate_hydrates_and_calcium_hydroxide_through_molecular_dynamics-CSIR-SERC (accessed on 6 October 2022).
44. Du, J.; Bu, Y.; Shen, Z. Interfacial properties and nanostructural characteristics of epoxy resin in cement matrix. *Constr. Build. Mater.* **2018**, *164*, 103–112. [CrossRef]
45. Al-Ostaz, A.; Wu, W.; Cheng, D.; Song, D.A.; Associate, C.R.; Al-Ostaz, A. A molecular dynamics and microporomechanics study on the mechanical properties of major constituents of hydrated cement. *Compos. Part B Eng.* **2010**, *41*, 543–549. [CrossRef]
46. Hajilar, S.; Shafei, B. Nano-scale investigation of elastic properties of hydrated cement paste constituents using molecular dynamics simulations. *Comput. Mater. Sci.* **2015**, *101*, 216–226. [CrossRef]
47. Marc 2021.3—Online Help (HTML). Available online: https://help.hexagonmi.com/bundle/marc_2021.3/page/marc_main.htm (accessed on 7 October 2022).
48. MSC Marc-Mentat—Help Reference. Available online: https://simcompanion.hexagon.com/customers/s/article/msc-marc-mentat---help-reference-doc9250 (accessed on 7 October 2022).
49. Konsta-Gdoutos, M.S.; Metaxa, Z.S.; Shah, S.P. Multi-scale mechanical and fracture characteristics and early-age strain capacity of high performance carbon nanotube/cement nanocomposites. *Cem. Concr. Compos* **2010**, *32*, 110–115. [CrossRef]
50. Abu Al-Rub, R.K.; Ashour, A.I.; Tyson, B.M. On the aspect ratio effect of multi-walled carbon nanotube reinforcements on the mechanical properties of cementitious nanocomposites. *Constr. Build. Mater.* **2012**, *35*, 647–655. [CrossRef]
51. Manzur, T.; Yazdani, N.; Emon, A.B. Effect of Carbon Nanotube Size on Compressive Strengths of Nanotube Reinforced Cementitious Composites. *J. Mater.* **2014**, *2014*, 960984. [CrossRef]
52. Mohsen, M.O.; Alansari, M.; Taha, R.; Senouci, A.; Abutaqa, A. Impact of CNTs' treatment, length and weight fraction on ordinary concrete mechanical properties. *Constr. Build. Mater.* **2020**, *264*, 120698. [CrossRef]
53. Centro de Publicaciones; Ministerio de Fomento; Secretaría General Técnica. *Instrucción de Hormigon Estructural. EHE-08*, 5th ed.; Ministerio de Transportes, Movilidad y Agenda urbana: Madrid, Spain, 2011; p. 702.

materials

MDPI

Article

Edgewise Compression and Three-Point Bending Analyses of Repaired Composite Sandwich Panels

Ricardo J. B. Rocha [1], Marcelo F. S. F. de Moura [2,*] and Raul D. F. Moreira [1,2]

[1] INEGI—Instituto de Ciência e Inovação em Engenharia Mecânica e Engenharia Industrial, Rua Dr. Roberto Frias 400, 4200-465 Porto, Portugal; rjrocha@inegi.up.pt (R.J.B.R.); rmoreira@fe.up.pt (R.D.F.M.)
[2] Departamento de Engenharia Mecânica, Faculdade de Engenharia da Universidade do Porto, Rua Dr. Roberto Frias, 4200-465 Porto, Portugal
* Correspondence: mfmoura@fe.up.pt; Tel.: +351-225081715

Abstract: In this work, the fracture behaviour of repaired honeycomb/carbon–epoxy sandwich panels under edgewise compression and three-point bending loading was analysed. Assuming the occurrence of damage resulting from a complete perforation leading to an open hole, the followed repair strategy consists of plug filling the core hole and considering two scarf patches with an angle of 10° in order to repair the damaged skins. Experimental tests were performed on undamaged and repaired situations in order to address the alteration in the failure modes and assess the repair efficiency. It was observed that repair recovers a large part of the mechanical properties of the corresponding undamaged case. Additionally, a three-dimensional finite element analysis incorporating a mixed-mode I + II + III cohesive zone model was performed for the repaired cases. Cohesive elements were considered in the several critical regions prone to damage development. The failure modes and the resultant load–displacement curves obtained numerically were compared with the experimental ones. It was concluded that the numerical model is suitable for estimating the fracture behaviour of sandwich panel repairs.

Keywords: sandwich panel; scarf repair; edgewise compression; three-point bending; cohesive zone modelling

Citation: Rocha, R.J.B.; de Moura, M.F.S.F.; Moreira, R.D.F. Edgewise Compression and Three-Point Bending Analyses of Repaired Composite Sandwich Panels. *Materials* **2023**, *16*, 4249. https://doi.org/10.3390/ma16124249

Academic Editor: Aniello Riccio

Received: 8 May 2023
Revised: 29 May 2023
Accepted: 5 June 2023
Published: 8 June 2023

1. Introduction

The structural application of sandwich panels is increasing due to their appealing characteristics. In fact, these components provide high stiffness and strength alongside a low weight, which is a vital aspect nowadays, regarding energy saving in moving structures, e.g., transportation and wind industries [1,2]. Nevertheless, sandwich-panel-based structures are vulnerable to damage, particularly under impact loading [3,4]. Generally, damaged large panels are rejected and replaced, which contributes to an unwanted ecological footprint and significant economic costs [5]. In these circumstances, the repair of the damaged components is required in order to extend their durability. The goal is to ensure that a large part of the mechanical properties of a damage component can be recovered after its repair, thus becoming a sustainable engineering approach [6,7].

Some researchers have dedicated their attention to the subject of sandwich repairs. Several experimental and numerical works focus on the behaviour of repaired sandwich composites with a two-dimensional configuration [8–11]. In these cases, bonded joint geometries are assumed to allow a simpler two-dimensional analysis. However, three-dimensional studies are recommended since patch repairs are usually circular, giving rise to a complex stress state under general loading. Raju et al. [12] performed experimental studies on composite sandwich panel specimens of polyurethane foam core and an aramid honeycomb core type damaged by impact loading. They carried out four-point bending and edgewise compression tests on undamaged, damaged and repaired specimens. The authors concluded that the strength recoveries revealed similar trends (values around 90%), with bending recovery showing a slight advantage compared to compression. Liu et al. [13] studied bonded repair on sandwich panels with one-side-skin and full-depth damage penetrating to the core under edgewise compression. They developed a three-dimensional progressive damage finite element model in order to predict the ultimate load, damage evolution process and stress distributions in adhesives. They studied the influence of different repair parameters including repair materials, taper ratios and repair techniques (scarf and step). The authors state that the numerical model provides an appropriate tool for the mechanical behaviour prediction of repaired sandwich panels. Ghazali et al. [14] studied the static mechanical performance of the repaired sandwich panels of a carbon–epoxy composite and honeycomb core. They performed edgewise compression and four-point bending tests on pristine and repaired panels with a stepped-scarf circular patch. A three-dimensional finite element analysis considering the different mechanical behaviours of the skins, core and adhesive was carried out. The numerical results were found to be in good agreement with the experimental ones regarding stiffness and strength predictions. Taotao Zhang et al. [15] performed experimental and numerical analyses of the damage propagation and ultimate strength of undamaged, open-hole and repaired sandwich panels under edgewise compressive load. They observed that the compressive strength of the open-hole damaged plate was about 34% of the undamaged one, increasing up to 76% for the repaired case. The numerical model accounts for intralaminar, interlaminar and honeycomb damage and was found to provide results in agreement with experimental data, thus contributing to improvements in the design and analysis techniques used to complete the scarf patch repair of sandwich structures. In this context, the model was subsequently used in parametric studies to assess the influence of the scarf angle, ply sequence and different overlaps on the ultimate strength and stress distribution in scarf-repaired sandwich plates. Yang et al. [7] studied the influence of the scarf patch repair angle on the bending strength of a honeycomb sandwich. They performed experimental and numerical analyses to evaluate the bending strength recovery. The authors concluded that the optimal solution involves using an angle of 30:1, which enables the total bending strength recovery of the sandwich structure. They also developed a three-dimensional finite element model incorporating spring elements and concluded that the model reproduces the experimental trends well.

In this work, sandwich carbon-fibre-reinforced polymer (CFRP)/honeycomb panels repaired using bonded circular patch scarf repairs were analysed under edgewise compression and three-point bending loading. Depending on the impact energy, damage can affect the outer skin, outer skin and internal core, or even cause full perforation. In this work, the most detrimental situation (i.e., full perforation) is analysed. Experimental tests on undamaged and repaired panels were performed in order to assess the influence of repair on the stiffness and strength of the panels. Three-dimensional finite element analyses involving cohesive zone modelling were employed to simulate the behaviour of the repaired panels. The results obtained revealed the suitability of the proposed methodology regarding the design predictions of the repaired sandwich panels.

2. Experimental Work

A sandwich panel comprising two skins of CFRP DDCFX005 (Torayca FT300-40B and DYNEEMA® SK99) fabric, with a [(±45)/(±45)/(90, 0)/(±45)/(±45)] layup with a 1.35 mm thickness and a core of NOMEX® Honeycomb with a 10 mm thickness, was used in this study (Figure 1). The elastic properties of the laminate were experimentally determined and are listed in Table 1. The core elastic properties are given in Table 2 [16,17].

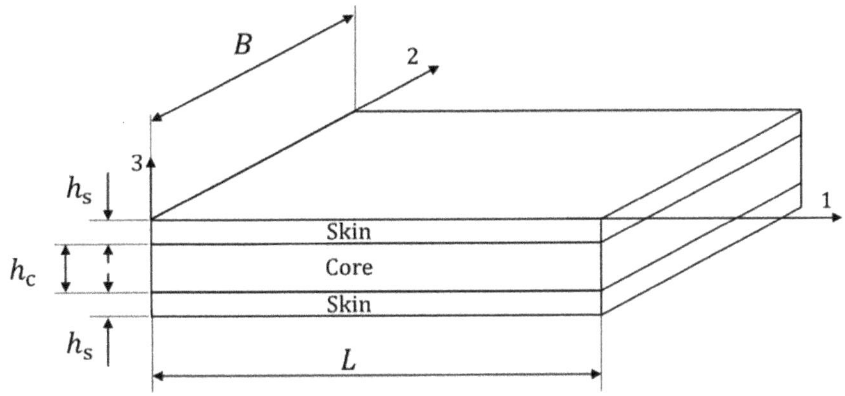

Figure 1. Schematic representation of the sandwich panels. Dimensions in mm: h_s = 1.35, h_c = 10; edgewise compression (L = 150, B = 100); three-point bending (L = 170, B = 250).

Table 1. Elastic properties of CFRP.

E_1 = 14,400 MPa	ν_{12} = 0.0017	G_{12} = 3550 MPa
E_2 = 14,400 MPa	ν_{13} = 0.04	G_{13} = 2070 MPa
E_3 = 2000 MPa	ν_{23} = 0.04	G_{23} = 2070 MPa

Table 2. Elastic properties of the NOMEX® Honeycomb [16,17].

E_1 = 0.45 MPa	ν_{12} = 0.9956	G_{12} = 0.11 MPa
E_2 = 0.45 MPa	ν_{13} = 0.0005	G_{13} = 38.62 MPa
E_3 = 258 MPa	ν_{23} = 0.0005	G_{23} = 63.12 MPa

The skins were bonded to the core using the structural epoxy adhesive ARALDITE® 2015-1 from Huntsman (E = 1850 MPa and ν = 0.3). Before bonding, the surfaces of the skin were sandpapered and subsequently cleaned with isopropyl alcohol to remove the impurities and contaminants of the bonding surface, thus improving the adhesion. The undamaged specimens were obtained by bonding the skins to the honeycomb core.

The damaged specimens were considered to suffer perforating damage induced by the high-velocity impact of a small projectile. In this context, a central hole was created, removing all the damaged region englobing the two skins and core. Additionally, two scarf chamfer profiles with an angle of 10° were machined on both skins using a CNC milling machine. Afterwards, a repair scheme was adopted with the aim of restoring a substantial part of the initial stiffness and strength. A cylindrical piece of core was bonded with ARALDITE® 2015-1 inside the core hole to replace the damaged and removed part of the core. Finally, two scarf patches manufactured with the same angle of 10° were bonded to both skins and to the core plug. It should be noted that both the parent laminate and patch were sandpapered (120-grit) in the scarf region to promote good adhesion. A constant adhesive thickness of 0.2 mm was assured using a calibrated wire located between the patches and the skins during the bonding process. The main characteristic of scarf repairs is the fact that patches are entirely inserted in the panel, thus not altering its aerodynamic behaviour (Figure 2).

Figure 2. Scarf repaired specimen.

Two different experiments were performed in this work: edgewise compression and three-point bending tests. For the edgewise compression tests, the procedure described in the ASTM D7137 standard was followed [18] and a special device with anti-buckling guides was used (Figure 3). The useful specimen dimensions are shown in Figure 4.

Figure 3. The experimental setup of the edgewise compression tests.

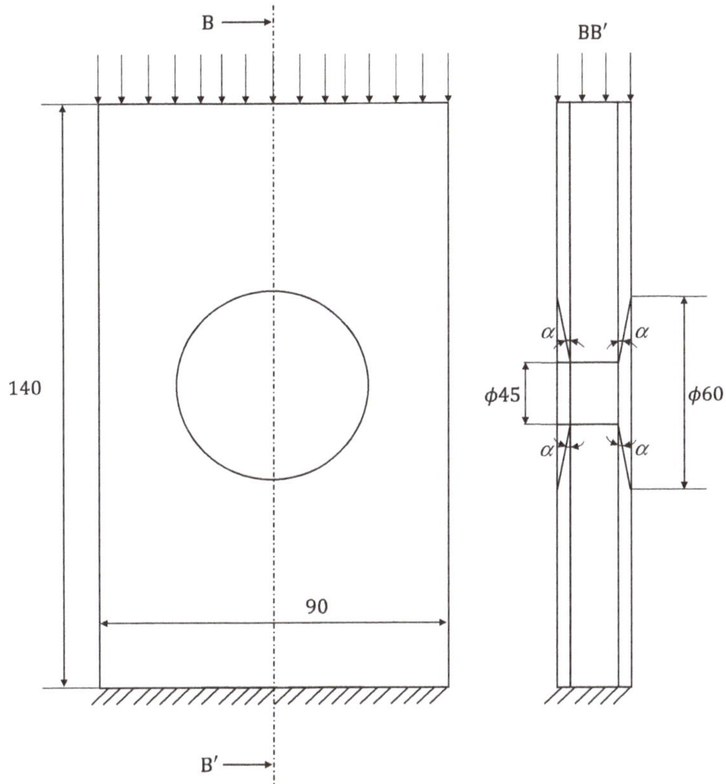

Figure 4. Schematic representation of the edgewise compression test (α = 10°).

For the three-point bending tests (Figure 5), an adaptation of the ASTM D790 standard was used [19]. Owing to the specimens' dimensions, a bigger damage region was considered (Figure 6) when compared to the compression tests, since previous studies have revealed that bending is less influenced by damage when compared to compressive behaviour.

Figure 5. The experimental setup of the three-point bending tests.

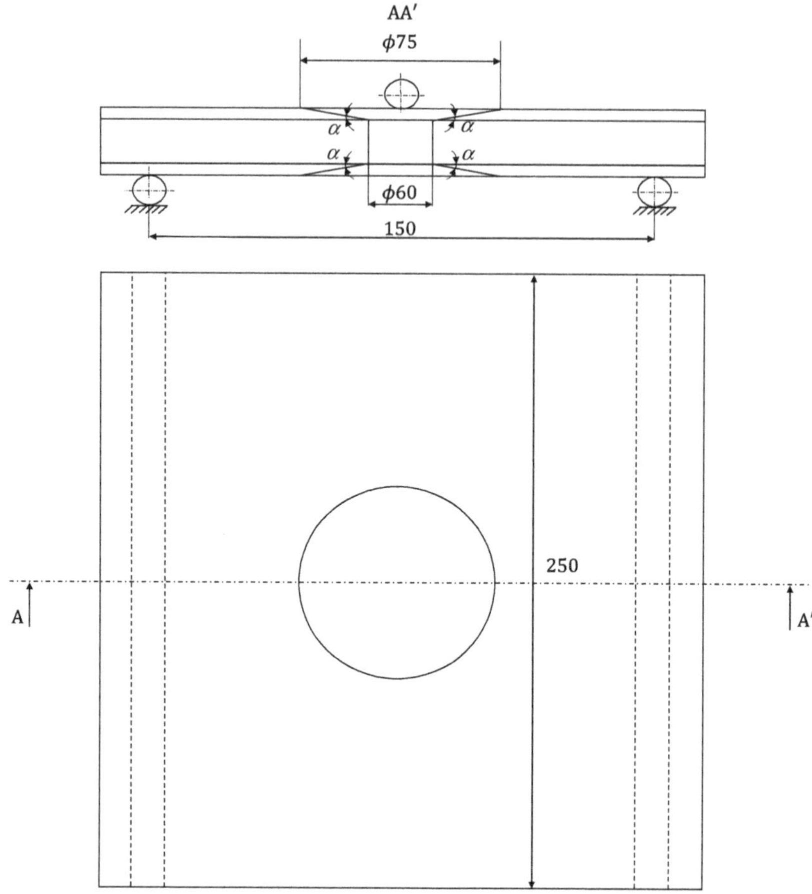

Figure 6. Schematic representation of the three-point bending test (α = 10°).

In both cases, the tests were performed under displacement control with a rate of 1 mm/min using a universal testing machine (INSTRON® 5900R) equipped with a load cell of 20 kN. The load–displacement (P-δ) curves were registered for subsequent analyses focusing on the evaluation of the initial stiffness and strength of the tested panels.

3. Numerical Analysis including CZM

Three-dimensional finite element analyses, including cohesive zone modelling, were performed considering the edgewise compression and three-point bending tests applied to the repaired panels. Owing to the development of complex loading at the repair region, a mixed-mode I + II + III damage model [20], considering the linear softening law, was used (Figure 7). The quadratic stress criterion is considered to deal with damage onset:

$$\left(\frac{\sigma_{\mathrm{I}}}{\sigma_{\mathrm{u,I}}}\right)^2 + \left(\frac{\sigma_{\mathrm{II}}}{\sigma_{\mathrm{u,II}}}\right)^2 + \left(\frac{\sigma_{\mathrm{III}}}{\sigma_{\mathrm{u,III}}}\right)^2 = 1 \tag{1}$$

where ($\sigma_{\rm I}$, $\sigma_{\rm II}$, $\sigma_{\rm III}$) are the mode I, II and III traction components, respectively, and ($\sigma_{\rm u,I}$, $\sigma_{\rm u,II}$, $\sigma_{\rm u,III}$) are the corresponding local strengths. Damage propagation was simulated considering the linear energetic criterion:

$$\left(\frac{G_{\rm I}}{G_{\rm Ic}}\right)+\left(\frac{G_{\rm II}}{G_{\rm IIc}}\right)+\left(\frac{G_{\rm III}}{G_{\rm IIIc}}\right)=1 \tag{2}$$

where G_i and G_{ic} (i = I, II, III) represent the strain energy release rate components and the corresponding critical values, respectively. After damage initiation (Equation (1)), a damage parameter (d) must be considered in order to mimic material softening:

$$\sigma_{\rm m}=(1-d)k\delta_{\rm m} \tag{3}$$

where k is the interfacial stiffness, $\delta_{\rm m}$ is the equivalent mixed-mode I + II + III relative displacement ($\delta_{\rm m}=\sqrt{\delta_{\rm I}^2+\delta_{\rm II}^2+\delta_{\rm III}^2}$), and $\sigma_{\rm m}$ is the corresponding mixed-mode I + II + III traction. After some algebraic manipulations [20], Equations (1) and (2) can be used to obtain the equivalent mixed-mode displacements at damage onset ($\delta_{\rm om}$) and at failure ($\delta_{\rm um}$), according to Figure 7. These parameters are used to define the damage parameter

$$d=\frac{\delta_{\rm um}(\delta_{\rm m}-\delta_{\rm om})}{\delta_{\rm m}(\delta_{\rm um}-\delta_{\rm om})} \tag{4}$$

that is used in Equation (3) to simulate material stiffness reduction.

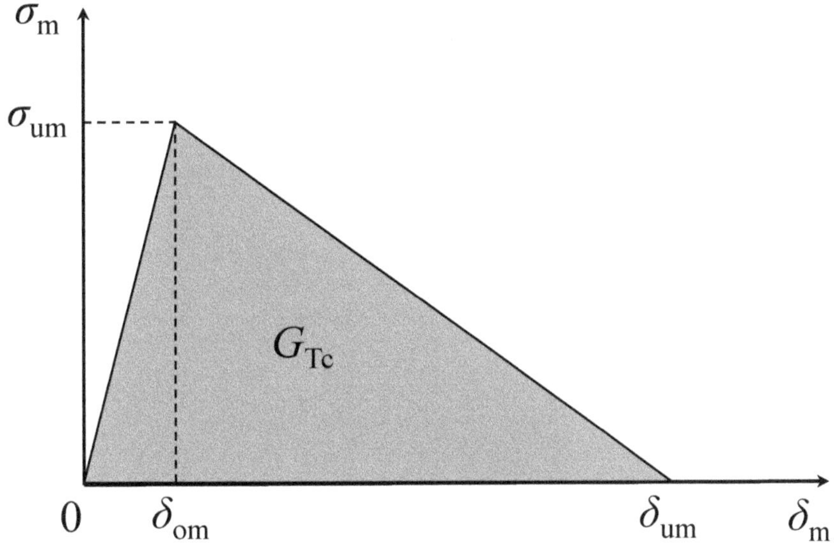

Figure 7. The linear softening cohesive law under mixed-mode I + II + III loading (subscript m): $\delta_{\rm om}$—damage onset relative displacement; $\delta_{\rm um}$—ultimate relative displacement; $\sigma_{\rm um}$—local strength; $G_{\rm Tc}$—fracture energy.

The cohesive zone (CZ) elements were located at the critical regions prone to damage development: skin/patch in the scarf region, patch/plug, and plug/core interface planes (Figure 8).

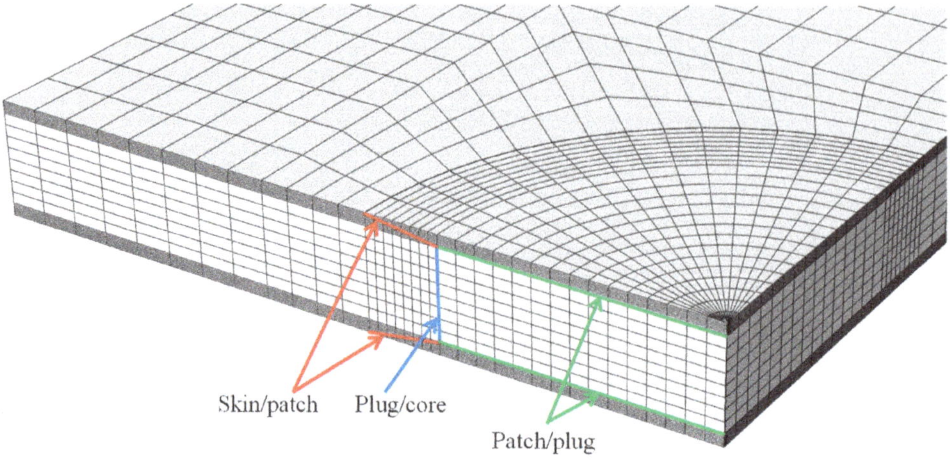

Figure 8. Location of cohesive zone elements.

4. Results

4.1. Edgewise Compression Tests

Experimental edgewise compression tests, considering an undamaged plate and four repaired ones, were performed. The undamaged case fails due to local crushing at its extremities (loaded and supported). The failure mode of the repaired specimens involves patch debonding followed by local crushing at the specimen mid-plane, which is caused by a reduction in the resistant section (Figure 9). The patch debonding reflects on a peak load that was assumed to be representative of the specimen strength.

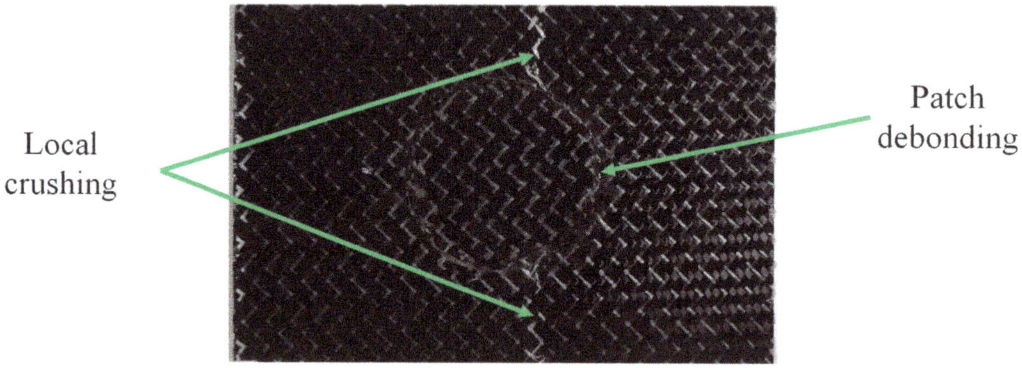

Figure 9. Typical damage obtained in edgewise compression tests of repaired specimens.

The load–displacement curves of the edgewise compression tests are presented in Figure 10. It can be observed that the initial stiffness and strength were not completely recovered after repair. Considering all the results, the initial stiffness and strength are in the range of 70–75% of the undamaged case. If we discard the lowest stiffness and strength case, assuming that it is an outlier result induced by any imperfection, the previous results change to the range of 75–80% of the undamaged case.

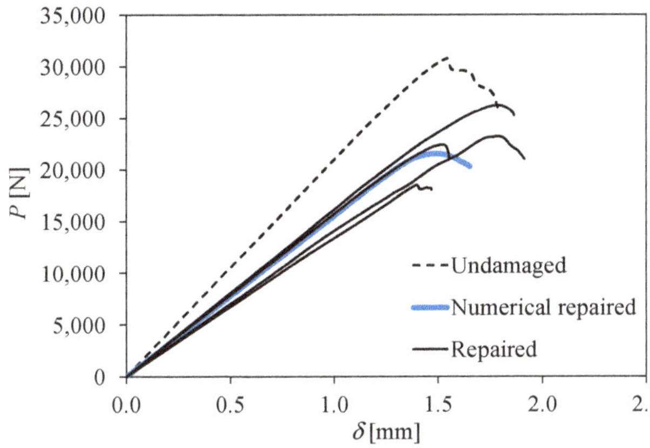

Figure 10. Load–displacement curves of the edgewise compression tests.

The three-dimensional numerical model considering the finite element analysis includes 22,016 solid elements (eight-node brick and six-node wedge) and compatible eight-node cohesive elements with null thickness. Only a quarter of the plate was simulated owing to symmetry conditions. A displacement (δ) was applied to the plate upper boundary to induce compressive loading, using small increments (0.02 mm per increment) in order to ensure stable damage development. The cohesive parameters used in the simulations were determined in previous works [16,17,21] and are listed in Tables 3 and 4.

Table 3. Cohesive parameters used for skin/core debonding analysis [16,17].

$\sigma_{u,I}$ **(MPa)**	$\sigma_{u,II}$ **(MPa)**	G_{Ic} **(N/mm)**	G_{IIc} **(N/mm)**
1.0	1.5	0.39	1.0

Table 4. Cohesive parameters used for scarf debonding analysis [22].

$\sigma_{u,I}$ **(MPa)**	$\sigma_{u,II}$ **(MPa)**	G_{Ic} **(N/mm)**	G_{IIc} **(N/mm)**
18.0	25.0	0.49	4.59

Figure 11 reveals the damage profile obtained numerically. As observed experimentally, patch debonding occurs, which defines the specimen strength. The numerical load–displacement curve is included in Figure 10, demonstrating that it represents the overall experimental trend well.

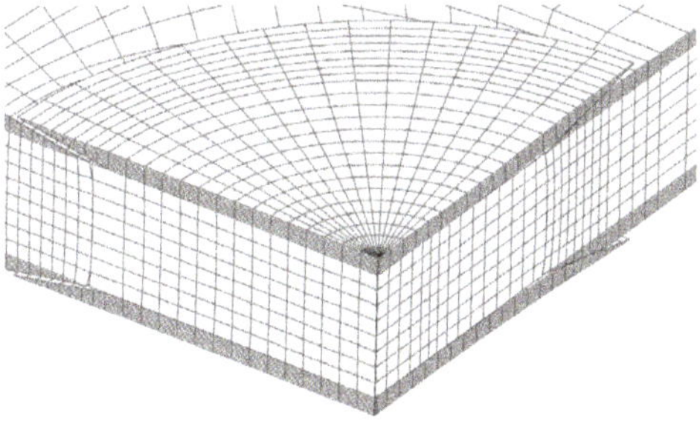

Figure 11. Damage obtained numerically in the edgewise compression tests.

4.2. Three-Point Bending Tests

Owing to premature failures, only two three-point bending measurements of the repaired specimens were obtained. In addition, both load–displacement curves are quite consistent with each other (Figure 12), which validates their consideration. In both cases, patch debonding of the external patch relative to the loaded surface (Figure 13) leads to a peak load that defines the specimen strength. Some non-linearity can be observed before the peak load, which can be explained by the development of non-negligible fracture process zones under mixed-mode loading. Similar to what happened in the edgewise compression tests, the stiffness and strength of the repaired plates are in the range of 75–80% of the undamaged case. In this type of test, the failure mode was dictated by the localised crushing of the upper skin at the specimen mid-span due to compressive stresses induced by bending.

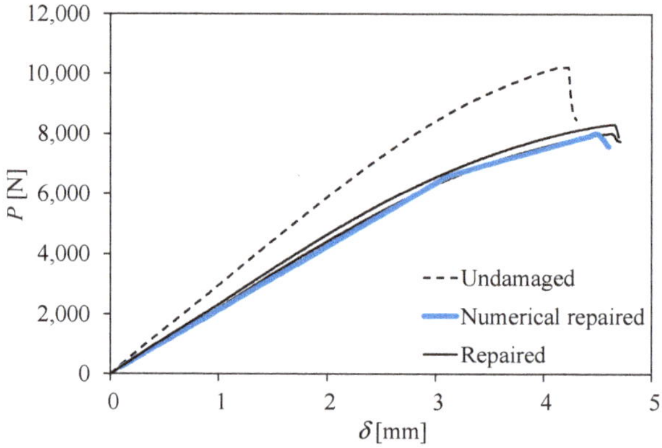

Figure 12. Load–displacement curves of the three-point bending tests.

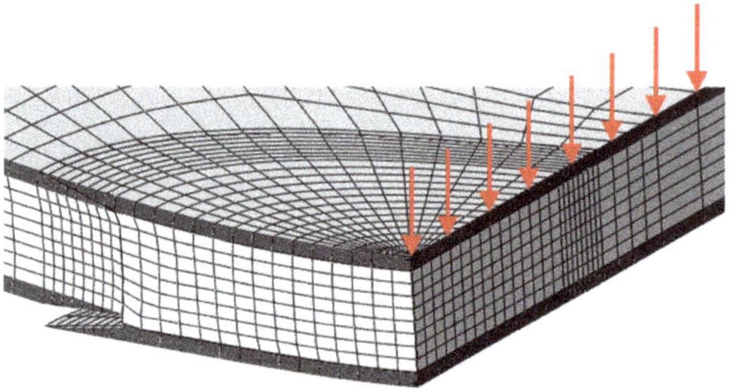

Figure 13. Damage obtained numerically in the three-point bending tests.

The numerical analysis was performed considering the same mesh and material properties used in the edgewise compression tests. The unique alterations consider the dimensions of the specimen and of the patch repair. A loading displacement (red arrows in Figure 13) was applied in the symmetry plane (specimen mid-span) by using small increments (0.02 mm) in order to avoid numerical instabilities. As observed experimentally, the failure mode occurs by the debonding of the external patch relative to the loaded surface, which allows the specimen strength to be defined. The numerical load–displacement curve reflects the experimental results well, revealing that the proposed model is adequate in order to simulate different repair strategies in three-dimensional progressive damage analysis.

5. Conclusions

In this work, the efficiency of scarf repairs on honeycomb/carbon–epoxy sandwich panels damaged by complete perforation (open-hole damage) was addressed experimentally and numerically under edgewise compression and three-point bending loading. The repair scheme is based on the plug filling of the damaged core with honeycomb and the use of two scarf patches with an angle of 10° in order to repair the damaged skins. In both loading cases, it was verified that the initial stiffness and strength of the corresponding undamaged situation were recovered up to 75–80% after repair. The main reason for this is the alteration of the failure mode which, in the repaired case, is mainly dictated by patch debonding that subsequently triggers the final collapse of the panel.

A three-dimensional numerical model based on finite element analysis, including mixed-mode I + II + III cohesive zone modelling, was developed for the repaired cases. Cohesive elements were considered in the critical regions prone to damage onset. The experimental failure modes were captured well by the model. The numerical load–displacement curves reproduce the observed experimental trends in both cases well. It was concluded that the model could be considered a useful tool regarding the optimization of the design of repairs for sandwich panels.

Author Contributions: Conceptualization, R.J.B.R., M.F.S.F.d.M. and R.D.F.M.; Methodology, R.J.B.R., M.F.S.F.d.M. and R.D.F.M.; Software, M.F.S.F.d.M. and R.D.F.M.; Validation, R.J.B.R., M.F.S.F.d.M. and R.D.F.M.; Formal analysis, R.J.B.R., M.F.S.F.d.M. and R.D.F.M.; Investigation, R.J.B.R., M.F.S.F.d.M. and R.D.F.M.; Writing—original draft, R.J.B.R., M.F.S.F.d.M. and R.D.F.M.; Writing—review & editing, M.F.S.F.d.M. and R.D.F.M.; Visualization, M.F.S.F.d.M. and R.D.F.M.; Supervision, M.F.S.F.d.M. All authors have read and agreed to the published version of the manuscript.

Funding: This research was funded by the "Fundo Europeu de Desenvolvimento Regional (FEDER)" for the financial support through the project, "Soluções avançadas para materiais de impacto, reparação de aeroestruturas em compósito e sua monitorização" (MOSHO) NORTE-01-0247-FEDER-033796.

Institutional Review Board Statement: Not applicable.

Informed Consent Statement: Not applicable.

Data Availability Statement: The data presented in this study are available on request from the corresponding author. The data are not publicly available due to privacy.

Conflicts of Interest: The authors declare no conflict of interest.

References

1. Katnam, K.B.; Comer, A.J.; Roy, D.; da Silva, L.F.M.; Young, T.M. Composite Repair in Wind Turbine Blades: An Overview. *J. Adhes.* **2015**, *91*, 113–139. [CrossRef]
2. Mishnaevsky, L. Repair of wind turbine blades: Review of methods and related computational mechanics problems. *Renew. Energy* **2019**, *140*, 828–839. [CrossRef]
3. Meo, M.; Vignjevic, R.; Marengo, G. The response of honeycomb sandwich panels under low-velocity impact loading. *Int. J. Mech. Sci.* **2005**, *47*, 1301–1325. [CrossRef]
4. Lacy, T.E.; Hwang, Y. Numerical modeling of impact-damaged sandwich composites subjected to compression-after-impact loading. *Compos. Struct.* **2003**, *61*, 115–128. [CrossRef]
5. Katnam, K.B.; da Silva, L.F.M.; Young, T.M. Bonded repair of composite aircraft structures: A review of scientific challenges and opportunities. *Prog. Aerosp. Sci.* **2013**, *61*, 26–42. [CrossRef]
6. Bleay, S.M.; Loader, C.B.; Hawyes, V.J.; Humberstone, L.; Curtis, P.T. A smart repair system for polymer matrix composites. *Compos. Part A Appl. Sci. Manuf.* **2001**, *32*, 1767–1776. [CrossRef]
7. Caminero, M.A.; Lopez-Pedrosa, M.; Pinna, C.; Soutis, C. Damage Assessment of Composite Structures Using Digital Image Correlation. *Appl. Compos. Mater.* **2014**, *21*, 91–106. [CrossRef]
8. Chen, J. Predicting progressive delamination of stiffened fibre-composite panel and repaired sandwich panel by decohesion models. *J. Thermoplast. Compos. Mater.* **2002**, *15*, 429–442. [CrossRef]
9. Mahdi, S.; Kinloch, A.J.; Matthews, F.L.; Crisfield, M.A. The static mechanical performance of repaired composite sandwich beams: Part I—Experimental characterization. *J. Sandw. Struct. Mater.* **2003**, *5*, 179–202. [CrossRef]
10. Ramantani, D.A.; Campilho, R.D.S.G.; de Moura, M.F.S.F.; Marques, A.T. Stress and failure analysis of repaired sandwich composite beams using a cohesive damage model. *J. Sandw. Struct. Mater.* **2010**, *12*, 369–390. [CrossRef]

11. Ghazali, E.; Dano, M.-L.; Gakwaya, A.; Amyot, C.-O. Mechanical performance of repaired sandwich panels: Experimental characterization and finite-element modelling. *J. Sandw. Struct. Mater.* **2019**, *21*, 1357–1378. [CrossRef]
12. Raju, M.; Reddy, C.R.; Swamy, M.R.N.; Giridhar, G. Repair Effectiveness Studies on Impact Damaged Sandwich Composite Constructions. *J. Reinf. Plast. Compos.* **2006**, *25*, 5–16. [CrossRef]
13. Liu, S.; Guan, Z.; Guo, X.; Sun, K.; Kong, J. Edgewise compressive performance of repaired composite sandwich panels—Experiment and finite element analysis. *J. Reinf. Plast. Compos.* **2013**, *32*, 1331–1347. [CrossRef]
14. Ghazali, E.; Dano, M.-L.; Gakwaya, A.; Amyot, C.-O. Experimental and numerical studies of stepped-scarf circular repairs in composite sandwich panels. *Int. J. Adhes. Adhes.* **2018**, *82*, 41–49. [CrossRef]
15. Zhang, T.; Yan, Y.; Jin, C. Experimental and Numerical Investigations of Honeycomb Sandwich Composite Panels with Open-hole Damage and Scarf Repair Subjected to Compressive Loads. *J. Adhes.* **2016**, *92*, 380–401. [CrossRef]
16. de Moura, M.F.S.F.; Moreira, R.D.F.; Rocha, R.J.B.; Oliveira, C.F.M. Determination of the fracture energy under mode I loading of a honeycomb/carbon-epoxy sandwich panel using the asymmetric double cantilever beam test. *J. Sandw. Struct. Mater.* **2022**, *24*, 1977–1992. [CrossRef]
17. Moreira, R.D.F.; de Moura, M.F.S.F.; Rocha, R.J.B.; Oliveira, C.F.M. Mode II fracture characterisation of a honeycomb/carbon-epoxy sandwich panel using the asymmetric end-notched flexure test. *J. Sandw. Struct. Mater.* **2022**, *24*, 2030–2046. [CrossRef]
18. *ASTM D7137/D7137M-17*; Standard Test Method for Compressive Residual Strength Properties of Damaged Polymer Matrix Composite Plates. ASTM International: West Conshohocken, PA, USA, 2017. [CrossRef]
19. *ASTM D790-17*; Standard Test Methods for Flexural Properties of Unreinforced and Reinforced Plastics and Electrical Insulating Materials. ASTM International: West Conshohocken, PA, USA, 2017. [CrossRef]
20. Durão, L.M.P.; de Moura, M.F.S.F.; Marques, A.T. Numerical simulation of the drilling process on carbon/epoxy composite laminates. *Compos. Part A Appl. Sci. Manuf.* **2006**, *37*, 1325–1333. [CrossRef]
21. Oliveira, C.F.M.; de Moura, M.F.S.F.; Moreira, R.D.F.; Rocha, R.J.B. Mixed-mode I+II fracture characterisation of a honeycomb/carbon-epoxy sandwich panel using the Asymmetric Single-Leg Bending test. *Theor. Appl. Fract. Mech.* **2023**, *123*, 103725. [CrossRef]
22. Moreira, R.D.F.; de Moura, M.F.S.F.; Silva, F.G.A.; Reis, J.P. High-cycle fatigue analysis of adhesively bonded composite scarf repairs. *Compos. Part B Eng.* **2020**, *190*, 107900. [CrossRef]

materials

MDPI

Article

Predicting Failure of Additively Manufactured Specimens with Holes

Gina Eileen Chiara Schmeier [1,2], Clara Tröger [1,2], Young W. Kwon [1,*] and Delf Sachau [2]

1 Department of Mechanical & Aerospace Engineering, Naval Postgraduate School, Monterey, CA 93943, USA
2 Fakultät Maschinenbau, Helmut-Schmidt-Universität/Universität der Bundeswehr, 22043 Hamburg, Germany
* Correspondence: ywkwon@nps.edu

Abstract: Experimental and computational studies were conducted to predict failure loads of specimens containing different-sized holes made using the additive manufacturing (AM) technique. Two different types of test specimens were prepared. Flat specimens, manufactured from polylactic acid (PLA), were subjected to uniaxial loading. Tubular specimens, made of polycarbonate (PC), were subjected to combined loading that was applied using uniaxial testing equipment. Test specimens were uniquely designed and printed to apply the combined bending and torsional loads to tubular specimens. A newly developed failure theory was applied to predict the loads that would result in the fracture of these test specimens. This theory is composed of two conditions related to stress and the stress gradient to be simultaneously satisfied to predict failure. The failure loads predicted using the new failure criteria were compared closely with the experimental data for all test specimens. In addition, a semi-empirical equation was developed to predict the critical failure surface energy for different printing angles. The critical failure surface energy is a material property and is used for the stress gradient condition. Using the semi-empirically determined values for the failure criterion provided close agreement with experimental results.

Keywords: additive manufacturing; perforated specimen; failure load; combined loading; failure criterion

Citation: Schmeier, G.E.C.; Tröger, C.; Kwon, Y.W.; Sachau, D. Predicting Failure of Additively Manufactured Specimens with Holes. *Materials* **2023**, *16*, 2293. https://doi.org/10.3390/ma16062293

Academic Editors: Luís Miguel Pereira Durão and Robert Pederson

Received: 8 February 2023
Revised: 27 February 2023
Accepted: 9 March 2023
Published: 13 March 2023

1. Introduction

Additively manufactured, also known as 3D printed, parts are being used more frequently in a diverse set of applications. When the user-friendly technology of 3D printers is combined with polymer substrates, it became possible for an even greater body of researchers to produce parts for a wider variety of applications. As a result of this greater usage, understanding the behavior of parts produced in this way has become increasingly important. Compared to significantly heavier materials, polymer composites have a high stiffness and strength relative to their weight. This is a great advantage, especially for endeavors that depend on high-strength, low-weight materials. In the aerospace industry, for example, weight-saving components are increasingly being implemented using 3D printed polymers. Airbus relies on AM to reduce the weight of parts contained in the Racer high-speed helicopter [1].

Because the cost of 3D printers has dropped significantly in recent years, the technology has also become increasingly well known among private users who are inexpensively producing smaller models. A large variety of printable materials has also become accessible. Most printing is conducted using polymers such as polylactic acid (PLA), polycarbonate (PC), and acrylonitrile butadiene styrene (ABS), but metals that melt at low temperatures can also be used. Metals with a higher melting point and composites are also used as substrate materials, although they are currently only available at a very high price [2].

With the expansion of applications using AM, the question arises as to whether they are strong and durable enough to be used as load-carrying members. As a result, it is becoming increasingly important to understand the failure of these structures under various loading conditions. To address this question, research has been conducted to understand the strength and stiffness of 3D printed parts [3–12]. Those studies investigated the effect of various printing parameters on the strength and stiffness of 3D printed parts. Because most practical parts include geometric features that result in stress concentrations, this research focused on investigating the change in failure loading on samples containing a circular hole.

Having the ability to predict failure loads of such parts is necessary to better design 3D printed parts. Having a reliable theory that allows the estimation of failure before it occurs is a necessity. Many different failure criteria have been proposed, but they vary depending on whether the material of interest is isotropic or anisotropic, brittle, or ductile. Furthermore, different failure criteria have also been used depending on the shape of a defect such as notches in a part.

For specimens containing geometric features that induce a local stress concentration like a hole, a variety of failure criteria have been proposed [13–17]. Those failure criteria use a concept of characteristic critical distance. In other words, failures are predicted on the basis of stresses at critical distances or the average stresses up to critical distances. Different failure criteria have proposed different ways to determine the critical distances. More recently, a new unified failure criterion was developed [18–20]. The previous study demonstrated that the new failure criterion applies to brittle materials regardless of whether they contain a defect or not.

Even though there have been many studies on 3D printed parts, there has been no analytical attempt to predict the failure of 3D printed parts with notches, to the best of the authors' knowledge. Thus, the objective of this study was to experimentally measure failure loads of various 3D printed flat and tubular specimens containing different sizes of circular holes and to predict the failure loads using the new failure criterion. The flat specimens were made of PLA and subjected to uniaxial tensile loading, while tubular specimens were made of PC and subjected to combined loading. Because PLA specimens were stronger and stiffer along the printing direction, i.e., raster angle, they behaved as a fibrous composite [9,11]. As the raster angle varied, the PLA specimens behaved as laminated fibrous composites. One difference is that PLA specimens are made of one material while fibrous composites are made of fiber and matrix materials. This study evaluated how well the new failure criterion could predict failure loads for different kinds of 3D printed specimens subjected to uniaxial and combined loading, respectively.

To apply combined loading using a uniaxial testing machine, a new design of testing setup and test specimens was developed for this study, which took advantage of the flexibility of 3D printing. Combined loading can be applied by testing equipment that has a special testing machine to apply tensile and torsional loads simultaneously. To use a uniaxial testing machine for multiaxial loading, additional test setups were introduced. Those setups typically apply biaxial loading [21]. In this study, both bending and torsional loads were applied together as a combined load using uniaxial testing equipment.

The next section describes the preparation of 3D printed specimens and the design of combined loaded test specimens. Subsequent sections describe the material testing and an investigation of the failure criteria [18–20], followed by computational modeling results. Lastly, predictions of failure are made using the new failure criterion.

2. Preparation of 3D Printed Specimens

All the specimens were modeled using the program called Solidworks and converted into STL files using the default setting in Solidworks. Then, the specimens were fabricated and prepared using the fused filament fabrication (FFF) printing technique with different polymer filaments. PLA was selected for the flat, rectangular, test specimens because its properties can be changed by altering printing parameters, such as direction. An Ultimaker© S5 printer (Utrecht, Netherlands), which is shown in Figure 1, was used to print all of the specimens tested under this effort. The printer has a build dimension of

330 × 240 × 300 mm, and it has a range of resolution of 0.25 mm to 0.8 mm. Its nozzle temperature is between 180 °C and 280 °C. The nozzle can be heated up in less than 2 min.

Figure 1. Image of Ultimaker© S5 printer.

All the filaments used in this study were commercially available, and their material specifications are provided on their respective websites [22]. In this study, the PLA specimens were designed to produce orthotropic specimens with properties along the printing direction being measurably different from those measured in its orthogonal direction. To achieve that, the print settings shown in Table 1 were used to print all PLA samples. Every specimen was printed with the setting of 100% infill with a rectangular geometry measuring 140 mm long by 24 mm wide by 2 mm thick in the test section. As the specimens were held by the grips, the gauge length was set to be 100 mm as sketched in Figure 2. The width of specimens was chosen to fit the grip width of the test equipment, and the printing direction was varied from one sample to the next. Filament printing orientations were 0°, 90°, or +q/−q, where q was chosen to be 30°, 45°, or 60°. Figure 3 shows the different print directions measured relative to the direction of loading, with 0° corresponding to the axis along which the load was applied.

Table 1. PLA sample print settings.

Print Temperature	185 °C
Bed temperature	55 °C
Print speed	45 mm/s
Layer height	0.2 mm
Line width	0.35 mm

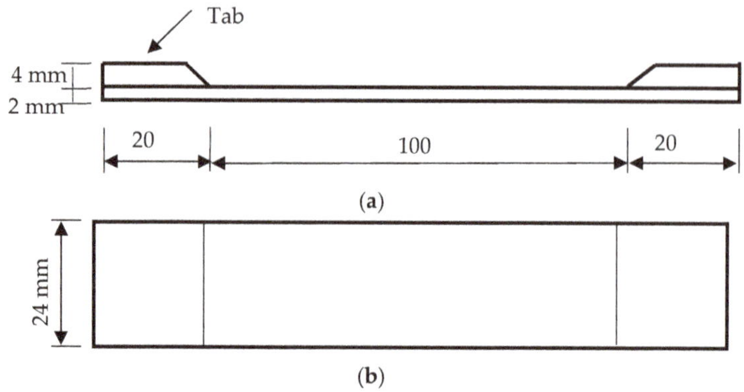

Figure 2. Rectangularly shaped test specimens: (**a**) side view; (**b**) top view.

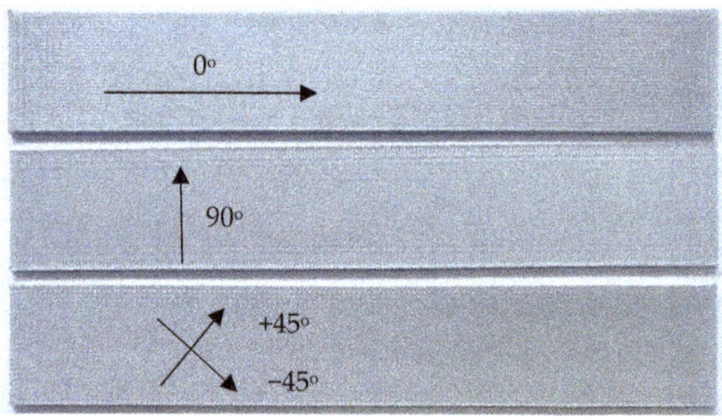

Figure 3. Test specimen different printing orientations.

Two types of PLA specimens were printed in this manner. One type of specimen was printed with a center hole while the other was printed without a hole. The latter specimens were used to determine the material properties of PLA specimens, while the former ones were used to measure how the failure load was influenced by such a geometric feature. Three to five specimens were printed and tested for every rectangular specimen to check the repeatability of test data. Because the test data were quite consistent, additional specimens were not printed and tested. The rectangular specimens without holes were printed with tabs for the grip sections of the specimen as one single piece as sketched in Figure 2. This was to ensure that failure occurred at the specimen midsection rather than at the grip sites. To simplify printing the samples, the tabs were printed on only one side of the specimen rather than both sides as that would have required the use of printing support material. Upon testing, this method was shown to not affect the overall test results. The specimens printed with center holes did not require tabs because a failure occurred at the minimum cross-section in the hole. The hole size was 3 mm, 6 mm, or 9 mm in diameter as shown in Figure 4. Three to five specimens of each type were printed to ensure enough duplicates for testing.

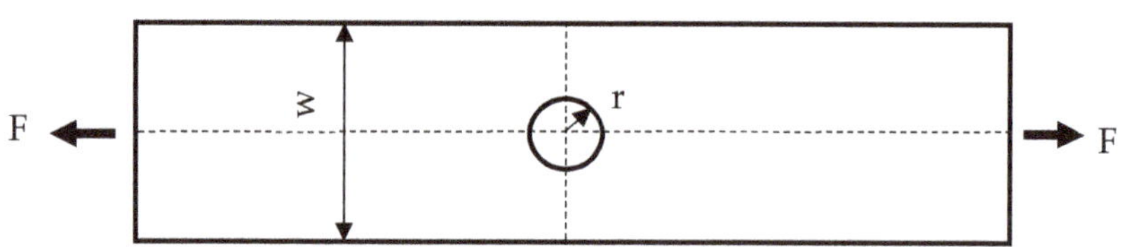

Figure 4. Center hole test specimen with 2 mm thickness (W = 24 mm, r = 1.5, 3, or 4.5 mm).

Two different techniques were used to generate the center hole. In one set of samples, it was printed in, while, in the other, the hole was drilled out. The former specimens were denoted by PHx.y (printed hole) while the latter specimens were called DHx.y (drilled hole), where x indicates the hole diameter, and y indicates the printing angle +q°/−q°. Figure 5 shows that there is a difference in quality along the edge of the hole based on the chosen manufacturing technique. Figure 5a shows the printed holes in the specimens using different print angles, while Figure 5b shows the drilled holes with the same orientations. Because the printed hole did not have a smooth circular edge, some portions of the edges were filed down to make them smooth. The drilling produced much smoother edges of the holes as compared in Figure 5. The specimens with drilled holes were modeled for predicting their failure stresses because smoother hole edges are much easier to model than rough edges.

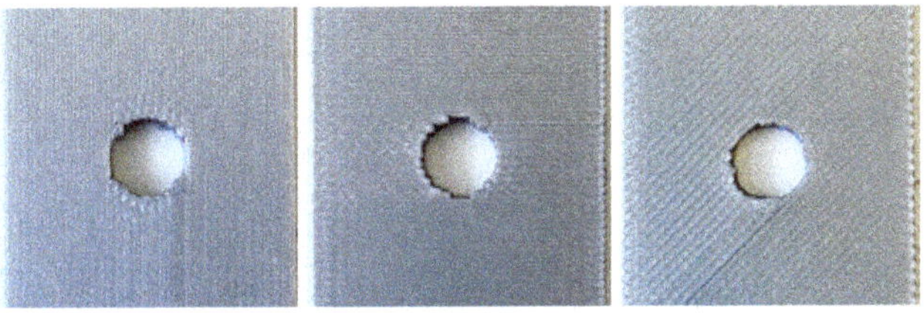

(**a**)

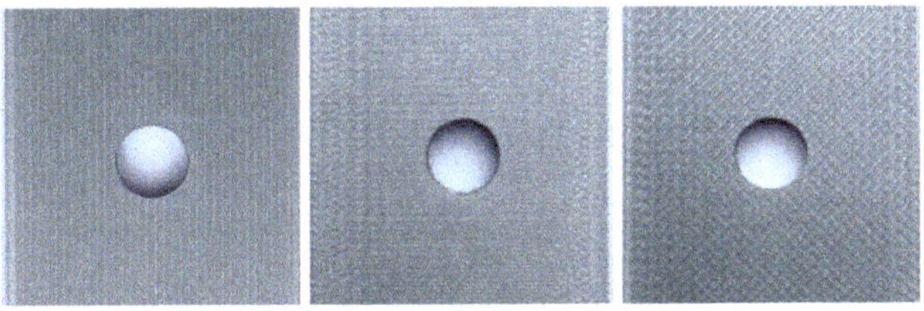

(**b**)

Figure 5. The 6 mm hole diameter samples: (**a**) PH6.0 (**left**), PH6.90 (**middle**), and PH6.45 (**right**); (**b**) DH6.0 (**left**), DH6.90 (**middle**), and DH6.45 (**right**).

All the tests were undertaken using an Instron© 2023 uniaxial test machine set at a crosshead speed of 2 mm/min. The first set of tests was performed on the rectangular specimens without a hole. Later, the samples with 0°, 90°, and ±45° raster angles were tested to determine their strength and stiffness. Some of those samples had strain gauges in the longitudinal and transverse directions to measure Poisson's ratio. Figure 6 is a post-test photograph of an instrumented test specimen. The tabs are not visible, however, because they are on the back side of the specimen.

Figure 6. Post-test photograph of a specimen without a circular hole.

Once the stiffness and strength of the PLA specimens were obtained for the specimens without holes, the specimens with holes were tested until fracture using the same testing condition and equipment. This showed a reduction in failure loads resulting from the holes, which is further discussed later.

3. Combined Loading Specimen Design

A new design was developed for conducting combined loading of test specimens using a standard uniaxial testing machine. The new designs were printed using AM because it affords researchers greater design flexibility than that offered using traditional machining methods.

Several different combinations of combined loading are possible. They might consist, for example, of biaxial loads, or a combination of tension and torsion among numerous other possibilities. In this study, the new test specimen was designed to simultaneously apply bending and torsion. The initial design of the combined loading specimen was a single-piece design as shown in Figure 7. To decrease printing times, parts of the design were hollow. The design consists of three portions: upper and lower handles, and a tubular test section. The lower handle is aligned with the testing section in the same plane while the upper handle is out of the plane of the test section. As the mount sections of both upper and lower handles are pulled apart in the uniaxial test machine, the test section, which may or may not have a hole, experiences both bending and torsion. If the upper handle is not out of the plane and aligned with the test section like the lower handle, the test section would be subjected to a bending load like the four-point bending test. The out-of-plane offset of the upper handle results in torsional loading.

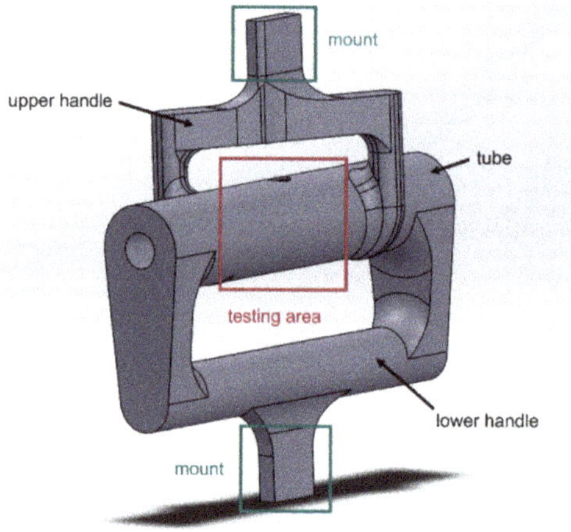

Figure 7. One-piece combined loading tubular specimen design.

To increase the torsion experienced by the sample, the offset distance of the upper handle was increased while the bending load was controlled by altering the distance between the upper and lower handles as seen in Figure 8. As the test specimen was subjected to axial loading F, the simplified theory was used to estimate resultant stresses. A numerical analysis using finite element analysis (FEA) was also conducted to determine the stresses experienced by the specimen. The bending stress at the outer surface of the test section is computed using Equation (1).

$$\sigma_b = \frac{M r_o}{I_b}, \tag{1}$$

where the bending moment M is calculated by

$$M = \frac{F L_b}{2}, \tag{2}$$

where L_b is the bending arm shown in Figure 8, and I_b is the second moment of inertia of the test cross-section, calculated as

$$I_b = \frac{\pi}{4}\left(r_o^4 - r_i^4\right), \tag{3}$$

where r_o and r_i are the outer and inner radii of the test section.

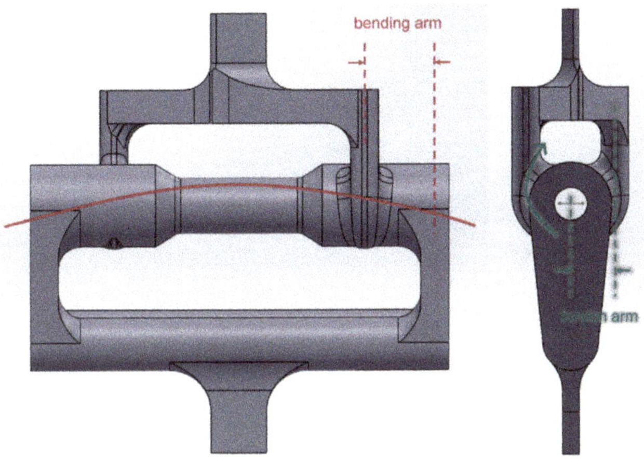

Figure 8. Initial combined bending and torsion design.

The shear stress resulting from the torsional loading is computed as

$$\tau_t = \frac{Tr_o}{J}. \tag{4}$$

The polar moment of inertia J is twice the value of the second moment of inertia I_b. The torsional moment T is calculated from

$$T = \frac{FL_t}{2}, \tag{5}$$

where L_t is the torsional arm as sketched in Figure 7. Lastly, the maximum normal stress of the combined stress state is expressed as

$$\sigma_{\max} = \frac{\sigma_b}{2} + \sqrt{\frac{\sigma_b^2}{4} + \tau_t^2}. \tag{6}$$

In a later section, these simplified analytical solutions are compared to the FEA results.

In the initial design, both the upper and the lower handles, as well as the test specimen, were printed as a single piece using a PC material. When tested, failure occurred at either the upper or the lower handle. Although FEA indicated that failure should occur at the test section because the handles were so much thicker and, hence, believed to be much stronger, this was not the case. The reason for such an unexpected failure was believed to be residual stresses in handle sections. In one instance of a one-piece specimen, a crack occurred in the lower handle during the 3D printing process because of residual thermal stress. In addition, printing the single-piece design presented other difficulties, e.g., taking longer and requiring more material.

To overcome the difficulties, the initial design was modified. The lower and upper handles, as well as the test specimens, were printed individually. The three parts of the modified design are shown in Figure 9. The lower and upper handles were printed out of an aluminum alloy to be deliberately stronger than the test section. Once both handles were printed, they could be used repeatedly unless major changes are required in the specimen design. The new design eliminated potential failure in the upper and lower handles guaranteeing failure at the test section of interest. Printing out test specimens in this manner also saved appreciable time and material.

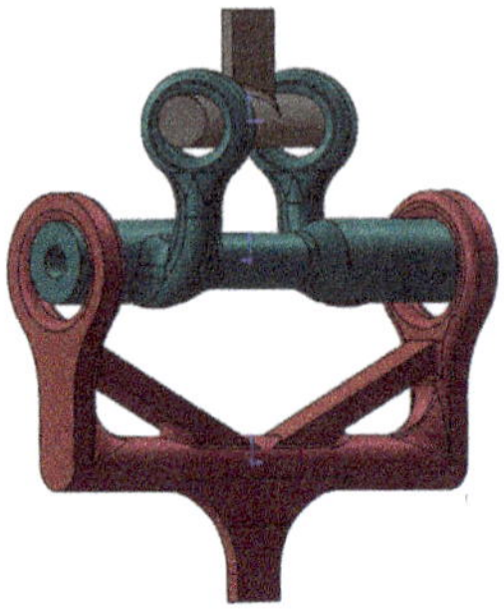

Figure 9. Three-part design: upper handle, test tube, and lower handle.

The test specimens for combined loading were printed using two different 3D printers. The Ultimaker© S5 was again used to print a black PC at 100% infill. These settings are listed below in Table 2. This setting was determined on the basis of the recommended temperature of the PC material by the manufacturer. The other printer was a Fortus© 450mc manufactured by Stratasys using a white PC. This printer is industrial AM equipment and can have a layer height from 0.127 mm to 0.330 mm with a part accuracy of ±0.0015 mm/mm. The default setting was used for the Fortus© 450mc printer. Using each printer, four different types of test specimens were fabricated as shown in Figure 10. Two of the test specimens had a longer torsional arm than the others. Of the two specimens with otherwise identical torsional arms, one had a 6 mm diameter hole and the other had no hole at all. The difference between the two 3D printers was the wall thickness of the test section. The test specimens made from Ultimaker© were 0.8 mm thick, while the test section manufactured using the Fortus© printer was 1.0 mm thick.

Table 2. Print settings for Ultimaker© printing PC.

Print Temperature	280 °C
Bed temperature	120 °C
Print speed	45 mm/s
Layer height	0.2 mm
Line width	0.4 mm
Brim line count	10

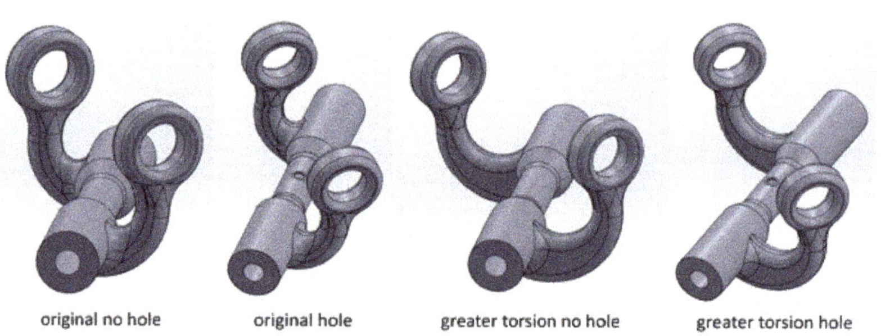

Figure 10. Different test specimens.

4. Testing and Results

All the tests were conducted using Instron following ASTM D638 as close as possible. Figure 11 shows a rectangular specimen with a hole installed at the grip of the testing equipment. Tensile tests of PLA specimens were first conducted on the specimens without any holes in the central section. The stress–strain curves of three test samples with different print angles are shown in Figure 12. The strength and stiffness of these specimens were different, depending on the printing angle. When the printing angle aligned with the loading direction, as it did in the case of the 0° specimen, the samples were the strongest and stiffest of any printing direction. This was a good indication that PLA specimens were orthotropic and possessed properties that have a strong correlation with the printing angle [9,11]. The longitudinal and transverse elastic moduli and tensile strength were obtained from the 0° and 90° specimens. The ±45° specimen was used to determine the shear modulus as described below.

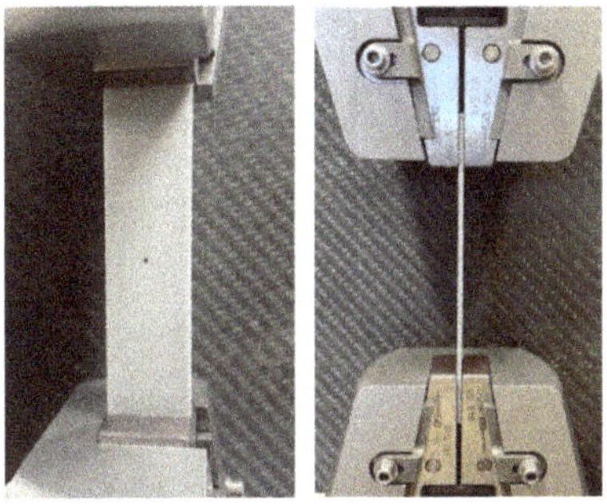

Figure 11. A rectangular test specimen with a hole held by the testing equipment.

The stress–strain constitutive matrix for an orthotropic material can be expressed as

$$\begin{Bmatrix} \varepsilon_1 \\ \varepsilon_2 \\ \gamma_{12} \end{Bmatrix} = \begin{bmatrix} S_{11} & S_{12} & 0 \\ S_{12} & S_{22} & 0 \\ 0 & 0 & S_{33} \end{bmatrix} \begin{Bmatrix} \sigma_1 \\ \sigma_2 \\ \tau_{12} \end{Bmatrix}, \quad (7)$$

where ε and γ are normal and shear strains, while σ and τ are the normal and shear stresses. The matrix coefficients S_{ij} are determined using the following expressions:

$$S_{11} = \frac{1}{E_1}, S_{12} = -\frac{\nu_{12}}{E_1} = -\frac{\nu_{21}}{E_2}, S_{22} = \frac{1}{E_2}, S_{66} = \frac{1}{G_{12}}, \quad (8)$$

where E_1 and E_2 were determined from the slope generated from test results of 0° and 90° specimens, respectively. Because the stress–strain curves have nonlinear sections, the initial part of the curves was used to determine the elastic moduli. Poisson's ratio ν_{ij} was also determined from the strain-gauge readings. Those material properties were also used later for computer modeling.

To determine the shear modulus G_{12}, a stress–strain transformation was used. When the printing angle is oriented at an angle q, the transformed equation becomes

$$\begin{Bmatrix} \bar{\varepsilon}_1 \\ \bar{\varepsilon}_2 \\ \bar{\gamma}_{12} \end{Bmatrix} = \begin{bmatrix} \bar{S}_{11} & \bar{S}_{12} & \bar{S}_{13} \\ \bar{S}_{12} & \bar{S}_{22} & \bar{S}_{23} \\ \bar{S}_{13} & \bar{S}_{23} & \bar{S}_{33} \end{bmatrix} \begin{Bmatrix} \bar{\sigma}_1 \\ \bar{\sigma}_2 \\ \bar{\tau}_{12} \end{Bmatrix}, \quad (9)$$

in which the superimposed bar denotes the values in the new coordinate system oriented with an angle q with respect to the original coordinate system. The first component in Equation (9) is expressed as

$$\bar{S}_{11} = S_{11}\cos^4\theta + (2S_{12} + S_{33})\sin^2\theta\cos^2\theta + S_{22}\sin^4\theta. \quad (10)$$

When q is ±45°, Equation (10) is simplified to

$$4\overline{S}_{11} = S_{11} + (2S_{12} + S_{33}) + S_{22}, \tag{11}$$

where $\overline{S}_{11}$ is obtained from the slope of the stress–strain curve of the ±45° specimen, and S_{11}, S_{22}, and S_{12} are obtained from 0° and 90° specimens. Thus, S_{33} is determined using Equation (11), and the shear modulus G_{12} is computed using Equation (8). All the material properties were measured and computed from PLA test specimens without holes. These are listed in Table 3 below.

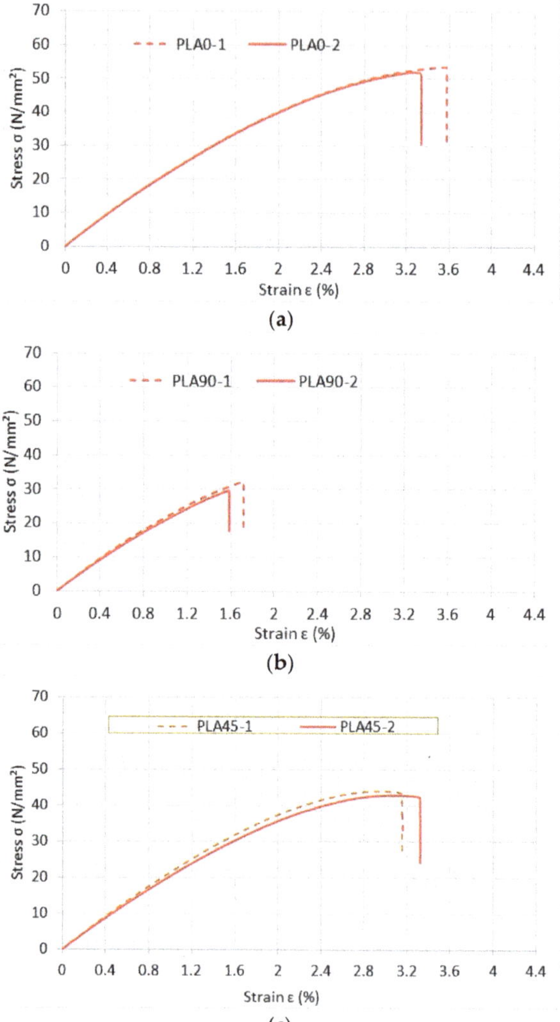

Figure 12. Stress–strain curves of PLA specimens with different printing angles: (**a**) 0°; (**b**) 90°; (**c**) ±45°.

Table 3. Properties of PLA specimens.

E_1	2.33 GPa
E_2	2.14 GPa
G_{12}	1.04 GPa
n_{12}	0.375
$(s_1)_{fail}$	57.7 MPa
$(s_2)_{fail}$	23.3 MPa

The next set of tests was conducted using perforated PLA specimens to compare printed holes to drilled ones. Figure 13 compares the stress–strain curves of both specimens with printed and drilled holes, both of which measured 6 mm in diameter. The printing angle was 0° for the specimens shown in Figure 13.

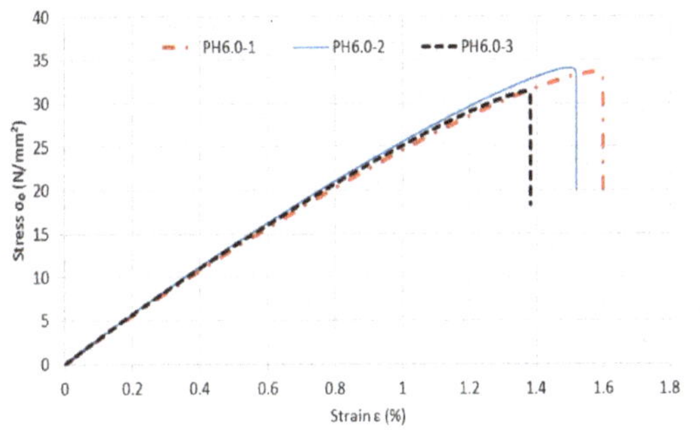

(**a**)

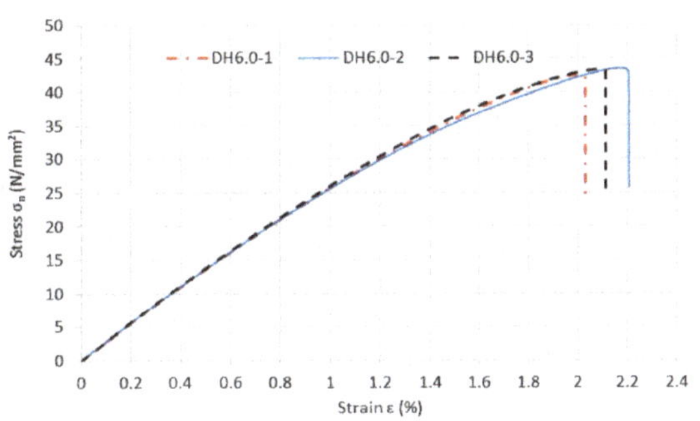

(**b**)

Figure 13. Comparison of stress–strain curves of specimens with (**a**) printed hole and (**b**) drilled hole with 6 mm diameter and 0° print angle.

The specimens with printed holes consistently exhibited lower failure stresses than those with drilled holes. Similar comparisons were also made to other specimens with different sizes of holes and printing angles. The specimens with drilled holes consistently exhibited greater strength than those with the printed ones. The main cause of this behavior was due to the smoother edges that the drilled holes had versus the jagged holes that occurred when they were printed as previously shown in Figure 5. Roughness at the edges of the printed holes resulted in a reduction in failure strength. It is important to point out that all the comparisons with analytical predictions were made using the drilled hole test specimens so that any hole roughness could be safely neglected in the FEA modeling.

PC specimens were tested in identical geometries using tabs just like the previous PLA samples. Initial testing was conducted on PC samples without holes to determine their strength and stiffness. The same three printing angles, 0°, 90°, and ±45°, were used and are plotted below in Figure 14.

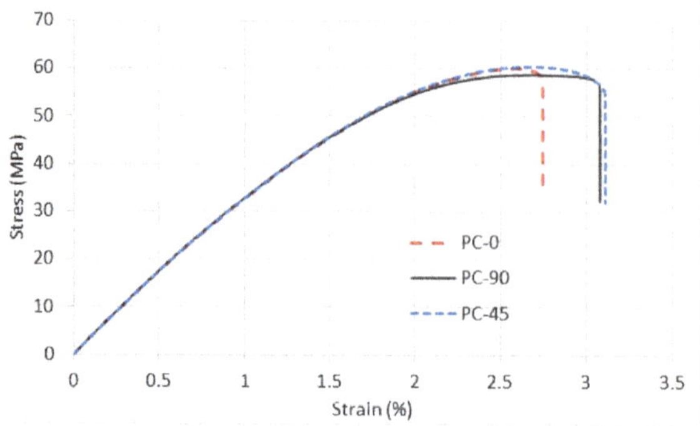

Figure 14. Stress–strain curve of PC specimens without a hole and different printing angles.

The stress–strain curves of these PC specimens were, however, almost identical regardless of which printing angle was used. This suggests that PC specimens should be considered as an isotropic material rather than an orthotropic one. The elastic modulus was 3.3 GPa, and a Poisson's ratio of 0.375 was obtained from the use of two perpendicular strain gauges.

5. Failure Criteria

The recently proposed unified failure criterion consists of two parts, both of which must be satisfied for failure to occur [18–20]. The failure criterion can be applied to specimens regardless of whether they have a defect, e.g., a crack or a hole. The first part is expressed as

$$\left|\sigma_{eff}\right| \geq \left|\sigma_{fail}\right|, \tag{12}$$

where σ_{eff} is the effective stress, and σ_{fail} is the failure strength of the material. For brittle materials, the maximum or minimum normal stress is used as the effective stress depending on tension or compression, and the failure strength is obtained from a uniaxial tensile or compressive test.

The second part is based on the stress gradient. The criterion is expressed as

$$\sigma_{eff} \geq \left(\left|\frac{d\sigma_{eff}}{ds}\right| E\kappa_{fail}\right)^{\frac{1}{3}}, \tag{13}$$

in which $\left|\frac{d\sigma_{eff}}{ds}\right|$ is the stress gradient and s is along the path of the failure. E is the elastic modulus, and κ_{fail} is another failure value that has the same units as the critical energy release rate.

To predict failure, both Equations (12) and (13) must be satisfied. Usually, one of the two dominates depending on the geometry. The two failure values, σ_{eff} and κ_{fail}, are needed to predict failure. Henceforth, the former is called failure strength while the latter is referred to as critical failure surface energy. To determine the failure strength and critical failure surface energy, unnotched and notched specimens were tested, and the respective values were measured. Then, those failure values were used for any specimen, regardless of sizes, shapes, or locations of notches, if it was the same material.

All the specimens with a circular hole in this study demonstrated that Equation (13) was the more limiting condition of the two. In other words, the effective stress at the initiation location was greater from Equation (13) than that from Equation (12). Therefore, the critical surface failure energy was used to predict failure stresses at notches. Because both PLA and PC are somewhat brittle, the maximum normal stress, or the largest principal stress, was used for the effective stress in both parts of the failure criterion. That is, the maximum normal stress and its gradient were computed at the failure location to predict failure load.

6. Numerical Modeling and Predictions

To predict failure at the edge of each hole, FEA using the Ansys program was conducted to determine the maximum normal stresses and their stress gradients. Rectangular PLA specimens were modeled only for one-quarter of the geometry because of the presence of two symmetric planes. The symmetric boundary conditions were applied to the symmetric edges (i.e., left and bottom edges), and a uniform displacement was applied to the top edge to resemble the actual testing condition. Figure 15 shows the mesh around a quarter circular hole based on a study of mesh sensitivity for converged values of the stress concentration factor. Both uniform and nonuniform meshes were considered in the study. A typical mesh had approximately 50,000 four-node quadrilateral elements and 40,000 nodes. Because failure occurred along the minimum cross-section of every rectangular specimen, the stress gradient was computed along the failure direction. All the analyses were linear elastic using the elastic moduli as discussed previously.

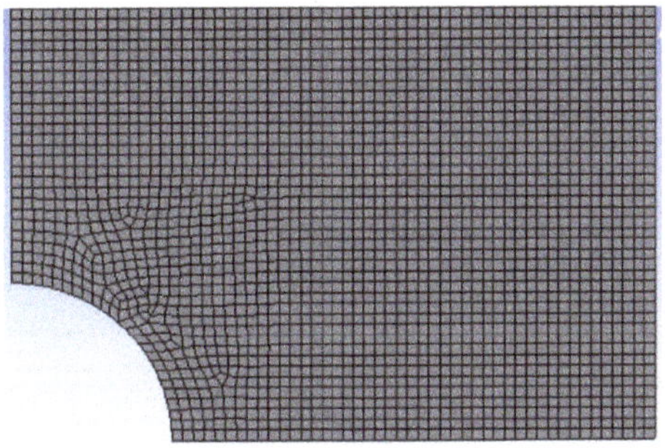

Figure 15. Finite element mesh around the hole.

As shown in Figure 13, the stress at which failure occurred was quite consistent for all specimens. Thus, the average value of the failure stresses was used for comparison to the analytical prediction. The failure stress in Figure 13 is the applied stress at the boundary where a uniform displacement was applied. It is not the stress at the edge of the hole. The comparison between experimental failure stress is compared to theoretical predictions, using the new two-part failure criterion, as shown below in Figure 16. Three different-sized holes were considered along with the three different printing angles. Only the drilled hole specimens were included because the rough edges would have been prohibitively difficult to model. Both failures compared very well with each other regardless of the hole size and printing angle. That is, the failure criterion [18–20] predicted applied failure stresses successfully for the PLA specimens.

Because PLA specimens had orthotropic material properties, different printing angles produced different material properties. That is, both stiffness and strength varied with the print angle. Thus, critical failure surface energy, κ_{fail}, also varied with the printing angle. This required determination of κ_{fail} for every different printing angle. Those values were used in Figure 16 to predict the applied failure stress. To minimize having to repeat tests to obtain κ_{fail} for each different angle, a semi-empirical equation was developed to predict the critical failure surface energy κ_{fail} as a function of print angle for PLA specimens.

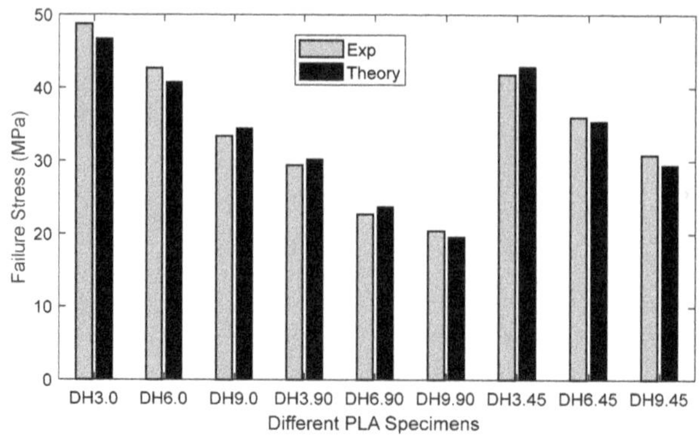

Figure 16. Experimental and theoretical failure stress for drilled PLA specimens.

The semi-empirical equation for the failure surface energy is

$$\left(\kappa_{fail}\right)_{\theta} = \frac{\left(\kappa_{fail}\right)_{0} + \left(\kappa_{fail}\right)_{\pi/2}}{2} + \frac{\left(\kappa_{fail}\right)_{0} - \left(\kappa_{fail}\right)_{\pi/2}}{2}(\cos 2\theta)^{7}, \quad (14)$$

where θ is the printing angle. Once the critical failure surface energy was determined using Equation (14), the applied failure stresses were again predicted for different printing angles to assess the semi-empirical equation. The comparison for the samples with 6 mm holes can be seen below in Figure 17. Both experimental and theoretical applied failure stresses were in close agreement, which confirmed that the semi-empirical equation produced reliable critical failure surface energy.

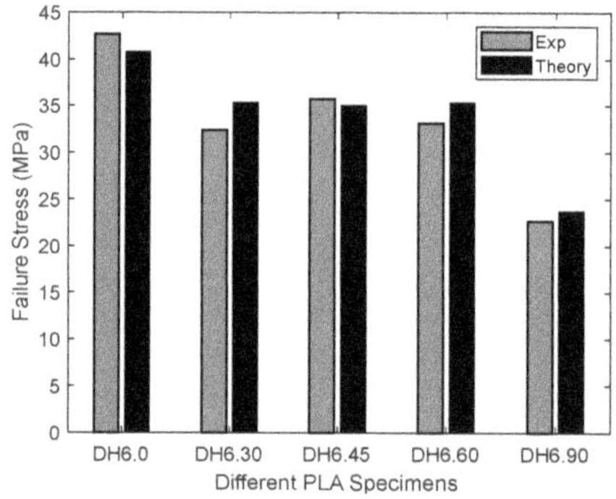

Figure 17. PLA failure stress for 6 mm drilled hole as a function of printing angle.

The next study was the combined loading of PC samples. These specimens were treated as isotropic due to the behavior exhibited previously and shown in Figure 14. A result from one of the combined loading scenario models is shown below in Figure 18. This FEA model included the whole specimens, as well as the upper and lower handles, as sketched in Figure 8. The bottom side of the lower handle was assumed to be fixed while the top side of the upper hand was loaded by external loading. Contact conditions were applied wherever the test specimen was in contact with the upper and lower handles. After a study of mesh sensitivity, about 200,000 eight-node solid elements were used based on the mesh sensitivity study.

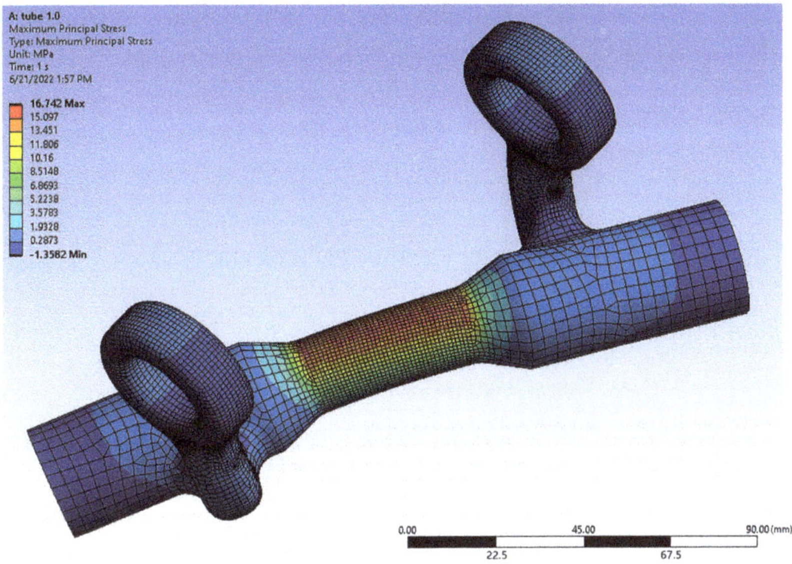

Figure 18. FEA of a combined loading specimen.

As expected, this test section achieved the highest stress state. Four different kinds of specimens, previously seen in Figure 10, were constructed using the Ultimaker© S5 and Fortus© 450mc printers, respectively. As a result, in total eight different specimens were printed because both 3D printers produced different thicknesses. In addition, the infill percentage for the Fortus© 450mc could not be controlled because it was already set for the printer. Every specimen had two duplicates because each took a long time to print.

The original specimens had a 35.5 mm bending arm and a 32.5 mm torsional arm. The greater torsion specimens had a 52.5 mm torsional arm, while the bending arm was unchanged. The test section was 40 mm long. These dimensions were the same for both specimens regardless of the printer used.

Figure 19 shows a combined loading specimen being tested in the uniaxial test machine. All the combined loading samples were tested until fracture. Figure 20 shows the fracture across the hole. As expected, the failure location agreed well with the FEA results.

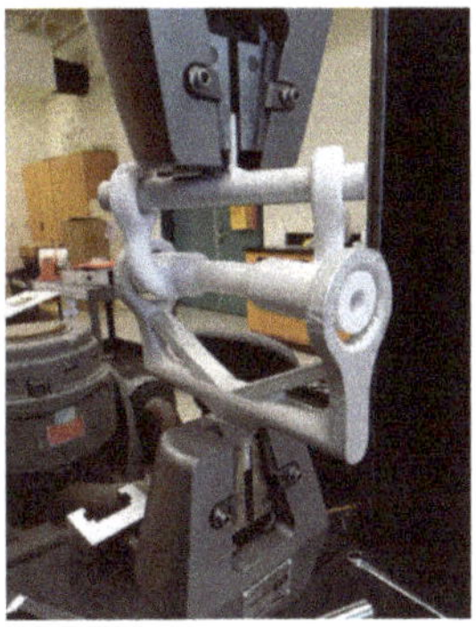

Figure 19. Combined loading specimen in the uniaxial test machine.

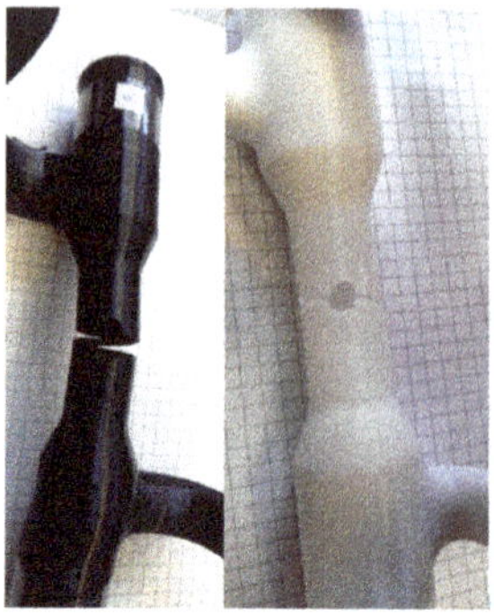

Figure 20. Post-test photograph of combined loading specimens (left specimen fabricated using Ultimaker© and right specimen fabricated using Fortus©).

Before comparing the experimental failure stresses to analytical values, FEA results were compared to the simplified stress analysis for the combined loading tests using specimens without holes using Equations (1)–(6). Table 4 compares the maximum normal stresses determined from FEA and the analytical solutions using Equations (1)–(6). The same failure loads were applied to FEA as well as Equations (1)–(6) to obtain the maximum normal stresses.

Table 4. Comparison of maximum normal stresses of tubular specimens subjected to combined loading.

	Thickness	σ_{max} (Analytic)	σ_{max} (FEA)	Error
Original	0.8 mm	44.18 MPa	46.85 MPa	6.04%
With no hole	1.0 mm	66.23 MPa	71.62 MPa	8.14%
Longer torsional arm	0.8 mm	49.20 MPa	54.70 MPa	11.19%
With no hole	1.0 mm	69.19 MPa	82.56 MPa	19.33%

The maximum normal stress varied from 6% to 20% between theory and FEA. One of the reasons for such differences was the torsional arms which were made of PC. Because the PC is more flexible than aluminum, the torsional arms deformed during the test. This would likely have affected the stress evolved in the test section and could have been one reason why the longer torsional arm resulted in a greater difference between the theory and FEA. In addition, torsional and bending arm lengths could not be precisely determined, as shown in Figure 8. Considering these differences, the analytical stresses were reasonable as compared to FEA.

Even though the analytical solutions are not very accurate as compared to the FEA solution, the analytical equations can be used in the designing process of the test setup and specimens without spending too much time on FEA.

Because there was no direct measurement of local stresses, failure stresses were computed using FEA as the experimental failure loads were applied to the FEA models. The failure stresses were the maximum normal stresses occurring in the specimens at the edge of the hole when the experimental failure loads were applied to the FEA model. To predict the failure stresses for the tubular specimens with a hole, Equations (12) and (13) were applied to every specimen. The stress gradient was computed using the FEA stresses at the two neighboring nodes along the failure direction starting from the edge of a hole. Then, an extrapolation technique was applied to find the stress gradient at the edge of the failure initiation site.

The results showed that Equation (13) is more critical than Equation (12). In other words, the failure stress from Equation (13) was greater than the failure stress from Equation (12). Because both parts of the criterion must be satisfied, the failure stress from Equation (13) was the predicted failure stress at the edge of the hole. The comparison between the experimental and analytical failure stress at the edge of the hole is shown in Figure 21. All the test specimens showed a good agreement between the theoretical and experimental failure stresses.

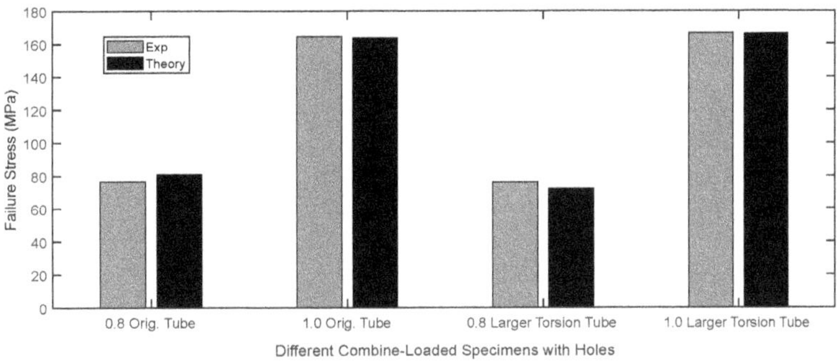

Figure 21. Comparison of the maximum stresses at the failure of combined loading specimens.

7. Summary and Discussion

Material properties of 3D printed PLA and PC specimens were experimentally measured using tensile testing. To avoid failure at the grip sections, test specimens used tabs at both ends. To simplify 3D printing, tabs were only printed on one side of each specimen so that no supporting material was needed. Print settings were chosen to induce orthotropic behavior in PLA specimens. Varying the angle of the printing direction using PLA resulted in behavior similar to that of a laminated fibrous composite. PC specimens, however, showed nearly isotropic behavior as the printing direction was changed.

Once obtaining material properties, a circular hole was introduced to PLA specimens which had different printing angles, while the overall specimen width, thickness, and length remained unchanged. These holes were made differently. In one scenario, the samples were printed with the hole; in the other, the hole was drilled out of the center of an otherwise pristine sample. The holes that were printed as part of the test specimens were rough and, as a result, induced a greater stress concentration. This manifested itself as a decreased failure load. Drilling the holes as the specimens were fully supported at their backsides, however, resulted in much smoother edges and a higher failure stress than that exhibited by the former samples. Using the samples with the printed holes was demonstrated to influence the failure load significantly; because of the lack of consistency and computational complexity in modeling the rough hole edges, they were not used in the study. Instead, the printed holes were considered for modeling and comparison.

In addition to flat PLA samples, tubular PC specimens were tested but under different loading conditions. These samples were subjected to combined loading using a specially designed test specimen geometry which could be tested using ordinary uniaxial test equipment. The test sample consisted of three unique parts, designed and fabricated using a 3D printing technique to produce complex shapes. The combined loading used to test these samples consisted of both bending and torsion. The magnitude of applied bending and torsional moments was controlled by varying the moment arm of each loading in the test setup. Stresses resulting from combined loading without stress concentrations were computed using the mechanics of materials approach, and the analytical stresses were in reasonable agreement with those calculated from FEA. The difference was due to the flexibility of the torsional arm of the tubular specimens, as well as an ambiguity in precisely defining the length of the moment arms. Both could be improved by modifying the design, although neither was attempted here since this was not the main focus of this study. Instead, the test setup was modeled using FEA to predict the resultant stresses occurring in the test specimens.

The failure loads of both PLA and PC specimens with a hole were predicted using a new failure criterion which used the stress gradient at the failure site. Maximum normal stresses were used for the failure criterion. The stress gradient was subsequently calculated using the slope of maximum normal stresses at the edge of the hole along the failure direction.

The PLA tensile specimens had three different hole diameters, while their width and length remained the same. In addition, the print angle was varied for the PLA specimens such as $+q^\circ/-q^\circ$, and q changed from 0° to 90° in fixed increments. There were nine different PLA specimens with a hole. At the loading boundary, the average applied stresses were calculated from the failure criterion and experimental data, and they were found to agree very well with each other.

The stress gradient-based criterion requires a material constant called the critical surface failure energy. When the printing angle was changed for the PLA samples, the critical failure surface energy was found to change as a result, similar to how laminated fibrous composites behave. To minimize redundant testing to determine the critical failure surface energy, a semi-empirical equation was developed to predict it for different printing angles. The failure stresses obtained using the critical failure surface energy predicted from the semi-empirical equation also agreed well with the experimental results. This confirmed that the semi-empirical equation was useful to estimate the critical failure surface energy.

The PC tubular specimens with a hole were also compared for their failure stresses between the analytical and experimental results. The failure stresses were the maximum normal stresses at the failure location of the holes as the specimens were subjected to combined loading which consisted of bending and torsion. For experimental results, the applied loads at failure were used in FEA to determine the maximum normal stresses at the failure locations because there was no direct measurement of local stresses during tests. There were four different tubular PC specimens with holes, with two different torsional arms printed using two different printers. The predicted failure stresses, i.e., maximum normal stresses, at the hole edges were in good agreement with the experimental data.

8. Concluding Remarks

A new failure criterion was used to predict failure loads and stresses for different types of 3D printed specimens with holes subjected to uniaxial and combined loading. To apply combined loading consisting of bending and torsion using uniaxial testing equipment, a novel test setup and testing specimens were designed and fabricated by taking advantage of flexibility in AM.

When failure stresses were calculated using these new criteria, they were found to be in good agreement with experimentally determined values for all the test specimens. As a result, this new criterion can be used to design a wide variety of structural parts with holes without the need for repetitive testing. This criterion was also useful for assessing the failure stress of 3D printed parts, which were printed in such a manner that their strength was found to vary as a function of printing angle.

Lastly, a semi-empirical equation was proposed to estimate the critical failure surface energy for different printing angles in parts, which is necessary to apply the new failure criterion for the prediction of failure loads and stresses. The equation was validated against additional experimental data. This equation can also help to eliminate additional tests to determine new material properties with different printing angles in designed parts.

Author Contributions: Y.W.K. was responsible for conceptualizing this study, providing guidance, and preparing the manuscript; G.E.C.S. and C.T. prepared all the specimens, conducted tests and FEA modeling, and analyzed the results; D.S. was responsible for supervising G.E.C.S. and C.T. during their studies, and he also reviewed the manuscript. All authors have read and agreed to the published version of the manuscript.

Funding: This research was funded by the Office of Naval Research for Young Kwon. The funding document numbers are N0001422WX00696 and N0001422WX00040.

Institutional Review Board Statement: Not applicable.

Informed Consent Statement: Not applicable.

Data Availability Statement: All the data are provided in this paper in the form of figures and tables.

Acknowledgments: The authors acknowledge the technical support by Chanman Park and proof-reading by Stanley Defisher.

Conflicts of Interest: The authors declare no conflict of interest.

References

1. Korner, R. CATEC Setzt auf 3D-Druck Beim Hinteren Rumpf des Airbus Helicopters RACER. Available online: https://www.3d-grenzenlos.de/magazin/kurznachrichten/airbushelicopters-racer-hinterer-rumpf-aus-3d-drucker-27828923/ (accessed on 24 June 2022).
2. Henry, S.; De Wever, L.; Vanhoorne, V.; De Beer, T.; Vervaet, C. Influence of Print Settings on the Critical Quality Attributes of Extrusion-Based 3D-Printed Caplets: A Quality-by-Design Approach. *Pharmaceutics* **2021**, *13*, 2068. [CrossRef] [PubMed]
3. Stava, O.; Vanek, J.; Benes, B.; Carr, N.; Měch, R. Stress relief: Improving structural strength of 3D printable objects. *ACM Trans. Graph.* **2012**, *31*, 48. [CrossRef]
4. Tanikella, N.G.; Wittbrodt, B.; Pearce, J.M. Tensile strength of commercial polymer materials for fused filament fabrication 3D printing. *Addit. Manuf.* **2017**, *15*, 40–47. [CrossRef]
5. Kovan, V.; Altan, G.; Topal, E.S. Effect of layer thickness and print orientation on strength of 3D printed and adhesively bonded single lap joints. *J. Mech. Sci. Technol.* **2017**, *31*, 2197–2201. [CrossRef]
6. Pyl, L.; Kalteremidou, K.-A.; Hemelrijck, D.V. Exploration of the design freedom of 3D printed continuous fiber-reinforced polymers in open-hole tensile strength tests. *Compos. Sci. Technol.* **2019**, *171*, 135–151. [CrossRef]
7. Dave, H.K.; Prajapati, A.R.; Rajpurohit, S.R.; Patadiya, N.H.; Raval, H.K. Open hole tensile testing of 3D printed parts using in-house fabricated PLA filament. *Rapid Prototyp. J.* **2020**, *26*, 21–31. [CrossRef]
8. Khosravani, M.R.; Reinicke, T. Mechanical strength of 3D-printed open hole polymer plates. *Procedia Struct. Integr.* **2022**, *41*, 664–669. [CrossRef]
9. Grant, A.; Regez, B.; Kocak, S.; Huber, J.D.; Mooers, A. Anisotropic properties of 3-D printed Poly Lactic Acid (PLA) and Acrylonitrile Butadiene Styrene (ABS) plastics. *Results Mater.* **2021**, *12*, 100227. [CrossRef]
10. Cuan-Urquizo, E.; Barocio, E.; Tejada-Ortigoza, V.; Pipes, R.B.; Rodriguez, C.A.; Roman-Flores, A. Characterization of the mechanical properties of FFF structures and materials: A Review on the experimental, computational and theoretical approaches. *Materials* **2019**, *12*, 895. [CrossRef] [PubMed]
11. Zohdi, N.; Yang, R. Material anisotropy in additively manufactured polymers and polymer composites: A review. *Polymers* **2021**, *13*, 3368. [CrossRef] [PubMed]
12. Horn, T.J.; Harrysson, O.L.A. Overview of current additive manufacturing technologies and selected applications. *Sci. Prog.* **2012**, *95*, 255–282. [CrossRef] [PubMed]
13. Whitney, J.M.; Nuismer, R.J. Stress fracture criteria for laminated composites containing stress concentrations. *J. Compos. Mater.* **1974**, *8*, 253–265. [CrossRef]
14. Taylor, D. Geometrical effects in fatigue: A unifying theoretical model. *Int. J. Fatigue.* **1999**, *21*, 413–420. [CrossRef]
15. Taylor, D. The theory of critical distances. *Eng. Fract. Mech.* **2008**, *75*, 1696–1705. [CrossRef]
16. Sapora, A.; Torabi, A.R.; Etesam, S.; Cornetti, P. Finite fracture mechanics crack initiation from a circular hole. *Fatigue Fract. Eng. Mater. Struct.* **2018**, *41*, 1627–1636. [CrossRef]
17. Braun, M.; Müller, A.M.; Milaković, A.-S.; Fricke, W.; Ehlers, S. Requirements for stress gradient-based fatigue assessment of notched structures according to theory of critical distance. *Fatigue Fract. Eng. Mater. Struct.* **2020**, *43*, 1541–1554. [CrossRef]
18. Kwon, Y.W. Revisiting failure of brittle materials. *J. Press. Vessel. Technol.* **2021**, *143*, 064503. [CrossRef]
19. Kwon, Y.W.; Diaz-Colon, C.; DeFisher, S. Failure Criteria for Brittle Notched Specimens. *J. Press. Vessel. Technol.* **2022**, *144*, 051506. [CrossRef]
20. Kwon, Y.W. Failure Prediction of Notched Composites Using Multiscale Approach. *Polymers* **2022**, *14*, 2481. [CrossRef] [PubMed]
21. Kalluri, S.; Bonacuse, P.J. (Eds.) *Multiaxial Fatigue and Deformation Testing Technique*; ASTM International: West Conshohocken, PA, USA, 1997.
22. The Widest Material Choice on the Market. Available online: https://ultimaker.com/materials (accessed on 22 February 2023).

materials

MDPI

Article

Damage Propagation by Cyclic Loading in Drilled Carbon/Epoxy Plates

Luis M. P. Durão [1,2,*], João E. Matos [1], Nuno C. Loureiro [3,4], José L. Esteves [4,5] and Susana C. F. Fernandes [1,2]

1 ISEP, Instituto Politécnico do Porto, Rua Dr. António Bernardino de Almeida, 4249-015 Porto, Portugal
2 Associate Laboratory for Energy, Transports and Aerospace (LAETA-INEGI), Rua Dr. Roberto Frias 400, 4200-465 Porto, Portugal
3 ISVOUGA, Instituto Superior de Entre Douro e Vouga, 4520-181 Santa Maria da Feira, Portugal
4 INEGI Instituto de Ciência e Inovação em Engenharia Mecânica e Engenharia Industrial, 4200-465 Porto, Portugal
5 FEUP Faculdade de Engenharia da Universidade do Porto, 4200-465 Porto, Portugal
* Correspondence: lmd@isep.ipp.pt

Abstract: Fiber reinforced composites are widely used in the production of parts for load bearing structures. It is generally recognized that composites can be affected both by monotonic and cyclic loading. For assembly purposes, drilling is needed, but holes can act as stress concentration notches, leading to damage propagation and failure. In this work, a batch of carbon/epoxy plates is drilled by different drill geometries, while thrust force is monitored and the hole's surrounding region is inspected. Based on radiographic images, the area and other features of the damaged region are computed for damage assessment. Finally, the specimens are subjected to Bearing Fatigue tests. Cyclic loading causes ovality of the holes and the loss of nearly 10% of the bearing net strength. These results can help to establish an association between the damaged region and the material's fatigue resistance, as larger damage extension and deformation by cyclic stress contribute to the loss of load carrying capacity of parts.

Keywords: bearing load; cyclic load; drilling damage; damage propagation

Citation: Durão, L.M.P.; Matos, J.E.; Loureiro, N.C.; Esteves, J.L.; Fernandes, S.C.F. Damage Propagation by Cyclic Loading in Drilled Carbon/Epoxy Plates. *Materials* **2023**, *16*, 2688. https://doi.org/10.3390/ma16072688

Academic Editor: Andrea Manes

Received: 28 February 2023
Revised: 14 March 2023
Accepted: 16 March 2023
Published: 28 March 2023

1. Introduction

Carbon/epoxy composite laminates are widely used in the production of load-bearing primary structures due to their highly specific strength, stiffness, fatigue strength and impact resistance, thermal conductivity, corrosion resistance, and good dimensional stability [1] when compared with metallic alloys. Examples of the advantageous use of fiber reinforced composites can be found in several fields of technology, see [2–5]. During the production process and further industrial application of parts made from carbon/epoxy composite laminates, drilling circular holes may be necessary to allow their assembly and incorporation into more complex products, using screws, bolts, pins, rivets or snap springs. Other machining operations can also be performed due to concerns related to dimensional accuracy and good surface quality requirements [1].

These operations are currently carried out by machining, and drilling is still the machining operation most widely used and studied in a considerable number of published papers covering several types of composites. Ghabezi and Khoran [6] investigated the influence of cutting speed, feed rate, and tool diameter in composite sandwich structures with a PVC core. Ghabezi et al. [7] carried out an experimental study to characterize the delamination and uncut fiber in the drilling of honeycomb structures. Yasar et al. [8] and Rajkumar et al. [9] also evaluated the influence of cutting parameters on hole quality for carbon/epoxy composites. A common outcome of these studies was the recognition of the superior contribution of feed rate to delamination extension. Most recently, Jinyang Xu et al. [10] conducted an experimental study on drilling damage for woven GFRP composites, contributing to a better understanding of the composites machining process. A recent review on the mechanisms of delamination formation, assessment, and suppression during the drilling of composites can be found in Geng et al. [11].

Even though drilling has been widely studied in recent years, for the purpose of drilling parameters optimization and their outcomes in terms of damage around the hole, there are few studies that address the fatigue behavior of composites in the vicinity of the hole, namely on the propagation of the damage when the parts are subjected to cyclic loads.

Generally, it is accepted that in the drilling process, delamination is the most frequent cause of damage and can reduce the load-bearing capacity of the parts. One of the main problems is related to the abrasive nature of the reinforcement fiber, which can lead to rapid tool wear and deterioration of the hole machined surfaces. The most typical damages caused by the drilling operation are pushout or exit delamination, peel-up or entry delamination, several hole wall damages such as fiber pullout, burrs, splintering, swelling, as well as thermal damages [1,12–14], see Figure 1.

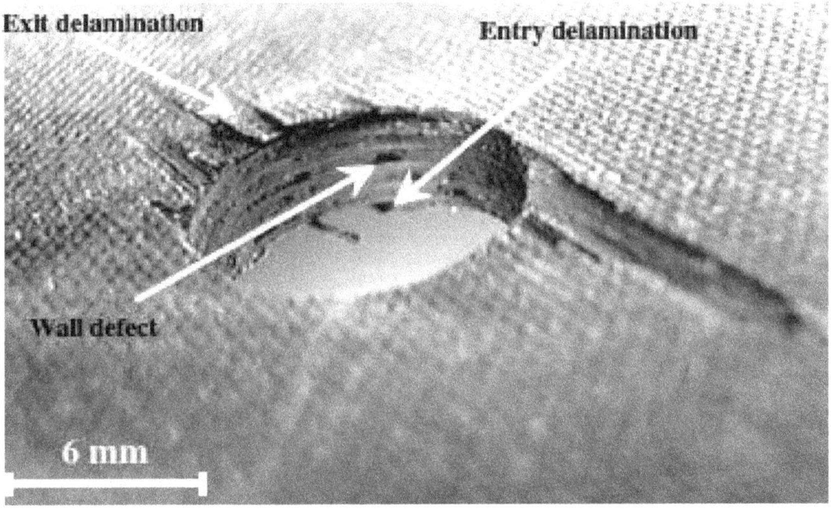

Figure 1. Drilling damages observed on CFRP laminate [1]. (Reprinted with permission from Ref. [1]. 2023, Springer Nature, https://www.springernature.com/gp).

Considering the drilling of a composite part, the desired results can be achieved with a smart combination of drilling tools and cutting parameters, tailored to the material to be cut, and the material properties such as the interlaminar fracture toughness in Mode I [15]. When using a common drill, there is a small region around the center of the chisel edge where the tool does not cut but extrudes the material, called the indentation zone. Considering that fiber-reinforced composites are, in general, more brittle than metals, it is unlikely that extrusion takes place, and orthogonal cutting could be assumed for the entire chisel edge. Drill bit geometry plays an important part in thrust force and, therefore, in delamination onset and propagation. The chisel edge of a twist drill acts like a punch over the uncut plies of the laminate. The consequence of this action is delamination whenever this thrust force exceeds the interlaminar resistance of the laminate. The size of the delamination zone has been proven to be related with the thrust force developed during the drilling process [16]. The use of special drill bits can contribute to the use of larger feeds without delamination onset. Overall, it is accepted that a drilling process that is able to avoid delamination should keep thrust forces below a threshold value, and this value is dependent on the uncut thickness, known as the critical thrust force (Fcrit) [17], see Equation (1). This critical thrust force is associated with several properties of the unidirectional laminate, such as the elastic modulus, E_1, the Poisson ratio, ν_{12}, the interlaminar fracture toughness in mode I, G_{Ic}, and the uncut plate thickness (h):

$$Fcrit = \pi \left[\frac{8 * G_{Ic} * E * h^3}{3\left(1 - \vartheta_{12}^2\right)} \right]^{\frac{1}{2}} \quad (1)$$

The work of Hocheng and Tsao has contributed significantly to the understanding of the delamination mechanism associated with diverse drilling conditions such as drill geometry [18] or tool wear. The hole can be predrilled to eliminate the thrust caused by the chisel edge. Thus, delamination risk is significantly reduced by the minimization of the thrust force, thus reducing delamination hazard [16]. Durão et al. [19–21] have followed this tool concept and developed several comparative studies comparing step and conventional drills. In [22], Tsao defined that an optimal ratio of first to second stage drill equal to 0.4 should be used to minimize delamination. Pilot hole drilling can be a good alternative for the reduction of thrust force, thus delamination. The tool geometry concept is essential to achieve good results, demonstrating the importance of a dedicated geometry for the drilling of fiber-reinforced laminates.

Studies on the load bearing consequence of drilling damage are usually concerned with the monotonic loading of drilled parts and their strength, as it was believed that these materials were not sensitive to cyclic effects. More recently, it has been recognized that composites can be affected by cyclic loading, leading to the need to establish the grounds for fatigue behavior knowledge. In fact, the initial drilling operation can act as a stress concentration notch, as it is understood that some delamination is prone to occur, followed by damage propagation from that point on, leading to multifaceted damage progression sequences and, finally, failure. This notch sensitivity can be affected by several factors including, among others, the laminate thickness, ply orientation, laminate size, notch size, and machining quality. This quality is governed by the drilling operation parameters—cutting speed and feed rate—as well as by the drill material and by the drill bit geometry.

The first known study on this effect is the work of Persson et al. [23], which shows the consequences of different tools on the fatigue behavior of drilled parts. From this starting point, several published papers are available that try to establish a sound correlation between the damage and fatigue behavior of composite plates. The first approach to this problem was from the aeronautical field [24,25], stating that, unlike metals, fatigue in composites develops in stages, each distinguished by a set of cracks. Experimental investigations into the damage development and residual strengths of open-hole specimens in fatigue were conducted and a model developed [26–28]. In this work, the authors used a cyclic load at 5 kHz to investigate the sequence of damage events leading to the final failure of the plate. It is interesting to note that fatigue delamination progression was assessed via X-ray CT. Finally, as previously expected, the log (number of cycles to failure) decreased linearly as the maximum fatigue stress level increased [26]. At the end of a predetermined number of cycles, a quasi-static loading until failure was performed [27]. Authors concluded that the residual strengths are similar for each of the severities tested Other more recent studies based on experimental analysis focused on evaluating the effect of a notch or stress concentrations on the fatigue performance of laminated structures, such as [29,30]. Muc and Romanowicz [30] demonstrated that crack initiation and growth are related to the interaction of the stress intensity factors. Hochard et al. [29] presented a fiber failure model for static and fatigue loads but recognized that further studies are needed. Yenigun and Kilickap [31] investigated the effect of drilling quality on the fatigue life of unidirectional, cross-ply and ±45° fiber angle carbon fiber-reinforced plastics. It was observed that with the decrease in drilling quality, a significant decrease occurred in the tensile force and fatigue life of the laminates. Sypt et al. [32] analyzed the failure mechanism of CFRP composite laminate under a cyclic pin-bearing load in quasi-isotropic laminates. In this thorough experimental and numerical modeling study, the authors presented the effect of several factors such as the loading factor and the loading ratio with a frequency of 5 Hz. Rui Guo et al. [33] analyzed the effect of tension-tension fatigue in composite rods with uniformly dispersed hybrid carbon/glass fiber, comparing the enhanced performance with glass fiber shell/carbon fiber core hybrid rods, showing that fatigue life has improved for the former mode.

Some studies also model the prediction of the open-hole strength plates, as seen in the papers of Camanho et al. [34] and Bao and Liu [35], or the studies on the effect of compressive stresses, such as in [36,37], thus providing a comprehensive base for the work here presented.

In the work here presented, two batches of carbon/epoxy plates were drilled under different machining conditions regarding the drill geometry, see Figure 2, and feed rate. The difference between the two baches is the stacking sequence, as one batch was of unidirectional plates $[0°]_{24}$ and the other was of cross-ply plates with the following stacking sequence $[(0/90)_6]_S$. During drilling, the thrust force was monitored to identify the likelihood of delamination onset. After drilling completion, the hole surrounding area was inspected by enhanced radiography, a non-destructive method combining a contrasting liquid penetrant with a digital image of the hole plus the surrounding delaminated edge. Radiographic images were obtained with the help of a 60 kV, 300 kHz Kodak 2100 X-ray system associated with a Kodak RVG 5100 digital acquisition system. From these images, the damaged area and other geometrical features of the damaged region were computed, using appropriate MatLab® tools, such as the Image Processing Toolbox.

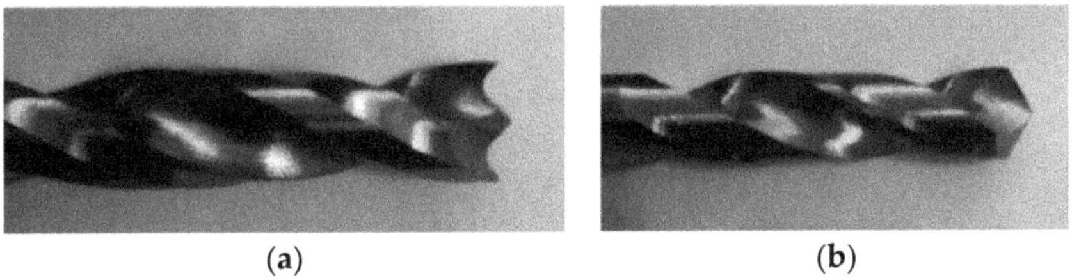

Figure 2. Drill geometries considered in experimental work: (**a**) Brad; (**b**) Twist.

Subsequently, the test specimens were subjected to cyclic bearing forces in a Shimadzu AG-X 100 kN Universal Testing Machine, (Shimadzu Corporation, Kyoto, Japan) as described in ASTM Bearing Fatigue Response of Polymer Matrix Composite Laminates—ASTM D6873—19 [38]. In the work here presented, specimens were not loaded until failure, as the test was halted after each thousandth cycle. Then, a new non-destructive inspection was carried out, and damage propagation extension was determined by comparing consecutive digital images by means of the image processing tools described above. This sequence was repeated twice for as many coupons as possible, providing that final failure did not occur, although failure did occur for a few. Finally, specimens that did not break under cyclic loading were bearing loaded until failure for residual strength assessment.

This experimental sequence has been presented in past papers from the same research team, see [39,40]. The only significant difference is the loading cycle, which is programmed with the help of the Trapezium software available with the universal testing machine.

The results of the experimental sequence here presented allow us to identify some differences in relation to the stacking sequences when under cyclic loads and, also, to have some information on the effects of cyclic loading on drilled composite plates. On the ither hand, cross-ply plates were able to withstand some deformation, showing some ovality of the hole, while most of the unidirectional plates failed after a reduced number of cycles. For the cross-ply plates, it was possible to draw a fatigue curve showing that a correlation between number of cycles and load amplitude can eventually be established. There is a decrease in the load bearing capacity of all the plates as the number of loading cycles increase. On the other hand, the load amplitude of the tests to unidirectional plates was always inferior to that of the cross-ply plates.

2. Materials and Methods

Fatigue testing is time and material consuming. The experimental work predicted needed to be adapted to keep the time frame within a reasonable period, as the main target was to have a significative number of results that could enhance further steps on the development of this type of material damage evaluation. Therefore, two batches of carbon/epoxy plates were produced: cross-ply and unidirectional. The test plates were produced from CIT (Composite Materials Italy, https://www.composite-materials.it/pagina.php?cod=1, accessed on 24 September 2022) carbon prepreg "CIT HS160 T700 ER450 UD tape 36%" and cured in an autoclave for one hour under 300 kPa and 130 °C, followed by cooling. The cross-ply test plates consisted of 24 layers of unidirectional prepreg, stacked in a symmetric cross-ply sequence, alternating 90° between each layer, corresponding to a stacking sequence of $[(0/90)_6]_s$, and coupons were supplied with the dimensions of $153 \times 35 \times 3.6\ mm^3$. The unidirectional plates had the same number of layers but all of them follow the same alignment, with all the fibers oriented towards the main load direction. Plate mechanical properties can be found in Table 1.

Table 1. Mechanical properties of prepreg CIT HS160 T700 ER450 UD tape 36% [41].

Property	Experimental Value	Unit	Standard
Ultimate tensile strength	1700	MPa	ASTM D3039
Elastic modulus	111	GPa	ASTM D3039
Poisson coefficient	0.29	—	ASTM D3039
Elongation	1.72	%	ASTM D3039
Interlaminar fracture toughness in Mode I	419	N/m	ASTM D5528

The experimental work began with the drilling of the laminate plates for thrust force monitoring, delamination measurement by enhanced radiography, automated computational algorithms of image processing and analysis and, finally, mechanical tests.

The drilling operation was performed in a HAAS VF-2 CNC, while thrust forces were monitored with the help of a Kistler 9171A load cell. As has been previously identified, feed rate is crucial compared to spindle speed in the development of thrust forces [42]. The cutting speed and feed rate were kept constant and equal to 2650 rpm and to 0.05 mm/rev, respectively. These cutting parameters were selected according to previous published works [20,42,43], keeping thrust forces below the value of the critical thrust force as well as the tool manufacturer's recommendation. When considering a layer thickness of 0.15 mm, F_{crit} is equal to 67 N, by Equation (1). A tool diameter of 6 mm was used combined with variations in drill geometry—Brad and twist (see Figure 2)—and material, as the latter were of tungsten carbide (WC) and of high-speed steel (HSS). BRAD drill is a commercial drill normally available in the tool manufacturers catalogue, and the twist drill is still the most common drilling tool geometry used in almost every tool shop. The standard option is a carbide drill, identified in this work as TWIST WC, but a TWIST HSS was also included with the intention to create a "bad" reference set, as normally this is considered a type of drill material to avoid when drilling of CFRPs is concerned.

After drilling, it is necessary to quantify the delamination extension caused by the machining operation. Normally, non-destructive tests (NDT) are used for this evaluation. For the damage assessment in this work, coupons were immersed in diiodomethane, a contrasting liquid, for 15 min and then radiographed with the help of a digital imaging system consisting of a 60 kV, 300 kHz Kodak 2100 X-ray system associated with a Kodak RVG 5100 digital acquisition system. The exposition time was set to 0.25 s. Once the digital images are acquired, it is possible to proceed with the necessary image processing to identify the damaged region and to compute the required parameters regarding the areas or diameters both the drilled hole and damaged area. The final purpose of the image processing and analysis sequence is the calculation of the assessment factor selected for this study, the Delamination factor (F_d), as in Bajpai et al. [44], Equation (2), where A_{max} is the maximum area (hole area + damage area) and A_{hole} is the geometrical value of a circle with the nominal diameter of the drill. An analogous experimental method was already implemented and used by the authors with interesting results [45].

$$F_d = \frac{A_{MAX}}{A_{hole}} \tag{2}$$

Further information on damage assessment criteria can be found in Geng et al. [11].

The next step was the completion of the mechanical tests. First, a few coupons were tested according to ASTM D5961-17 [46] Procedure A, both for unidirectional and cross-ply plates. This preliminary test was intended to set a strength value corresponding to the maximum bearing load, or 100%, that the material can bear in a quasi-static test. After this reference value was set, the plates were tested under different cyclic loadings at a low frequency, monitoring the mechanical response of the pin-loaded CRFP laminates until final failure of the plate or until a predefined number of cycles is reached. The test setup is in accordance with ASTM D6873 [38]. Due to the characteristics of the equipment used, it was decided to stop after 1000 cycles, corresponding to approximately 6 h of testing, or when pin displacement reached 0.6 mm, corresponding to 1/10 of the hole diameter. If failure occurs, the test was interrupted. Then, the plates were radiographed to observe and assess damage propagation caused by cyclic loading and then a new cyclic loading test was repeated. The maximum stresses used along this experimental phase are presented in Table 2, considering a loading ratio R = 0.1, as found in similar studies [26], to define the minimum stress. All the values of the fatigue testing sequence were rounded up for convenience.

Table 2. Maximum bearing stresses in cyclic tests (all values in MPa).

LOADING	Reference Value	90%	80%	70%	60%
UNIDIRECTIONAL	205	185	164	144	123
CROSS PLY	360	324	288	252	216
OBSERVATIONS	ASTM D5961	ASTM D6873			

Finally, all the plates without failure during cyclic tests were bearing tested according to ASTM D5961 [46] to check for residual strength. Results of this complete experimental sequence are presented in the following section.

3. Results and Discussion

3.1. Thrust Force Monitoring

The most significant value considered in this work was the axial thrust force (F_z) during drilling, as normally the main issue reported is the need to keep thrust force below the Critical Thrust Force (see Equation (1)) to avoid damage onset and propagation. Although the data collected include the forces in the plate plane (F_x, F_y), as well as torque, the range of values is too low to be regarded as influential in any outcome.

The results, presented for cross-ply plates only, include the typical development of thrust force curves along the drilling operation, see Table 3 and Figure 3. From the drilling data, it was also possible to compute the thrust force value when the uncut thickness was equal to 0.15 mm. This last result is comparable with the value of F_{crit} determined in the previous section.

Table 3. Thrust forces: maximum and at 0.15 mm from breakthrough (values in N).

DRILL GEOMETRY	MAX THRUST FORCE	THRUST FORCE AT 0.15 mm
BRAD WC	51	12.6
TWIST WC	87	12.5
TWIST HSS	183	22.1

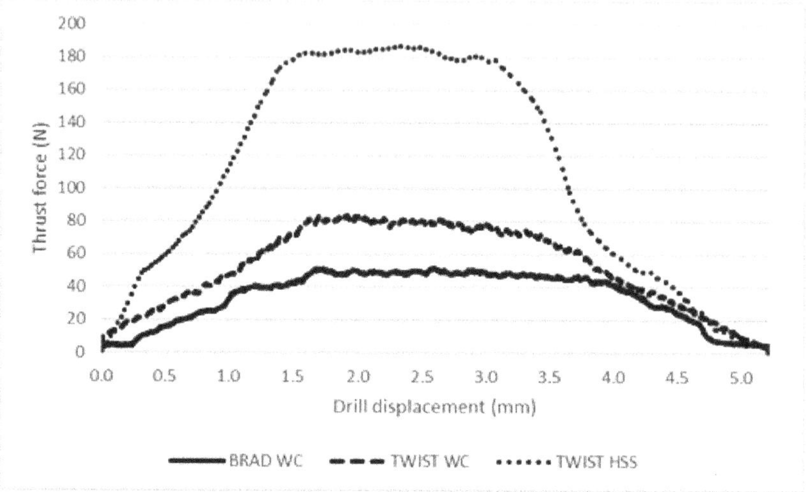

Figure 3. Comparison of thrust force during drilling of plates.

From the results available, it is not only possible to affirm that these curves are mainly influenced by drill geometry, but also that the maximum thrust force value or thrust force at a certain path registered during the drilling process is a function of drill geometry, with the plates being identical. Thus, when using the Brad drill, thrust force values were always lower than those observed with the Twist WC drill. The only possible reason for this is the drill geometry itself, as the twist drill has a higher piercing action exerted by the drill tip than the Brad drill design, which enables a pre-tensioning of the fibers before being cut. The result is a clean cut and a smoother cutting action, resulting in less roughness of the hole walls, an effect not included in this work, and, eventually, less delamination. When using Twist HSS drills, the thrust force values were always higher, confirming the established difficulty to machine CFRPs with HSS tools, leading to bad finishing and higher delamination around the machined hole.

3.2. Damage Assessment and Bearing Strength

After the drilling phase was completed, the next step was the radiography of all the plates drilled for a complete mapping of the damaged region around the hole. The results, considering the damage criteria defined in Equation (2), are presented in Table 4 as the average values for each drilling condition, including stacking sequence and drill geometry.

Table 4. Delamination factor (Equation (2)) and bearing strength of drilled plates.

PLATE	DRILL	DELAMINATION	BEARING STRENGTH (MPa)
UNIDIRECTIONAL	BRAD	1.12	191
	TWIST WC	1.12	210
	TWIST HSS	1.32	215
CROSS PLY	BRAD	1.18	386
	TWIST WC	1.12	380
	TWIST HSS	1.28	325

One of the motivations for this experimental work is that there should be some correlation between the damage extension, here assessed by the Delamination factor and the mechanical resistance of the plate that can be measured by the Bearing Strength. In Figure 4, the results of this correlation for the cross-ply plates are presented. Note that, even though there is some uncertainty in the results as is usual in composites, there is a clear trend on the cutback of the bearing strength as the damage extension is larger. Therefore, it is possible to affirm that a larger extension of the delaminated area, meaning a reduction in the consolidated resistant area in the vicinity of the hole, will cause a reduction in the bearing strength capacity of the plates. This outcome was expected and can be explained by the fact that a larger delamination extension will act as a pre-existing crack. As the crack is wider, the propagation of delamination makes the reduction in the cross-section resistance area easier.

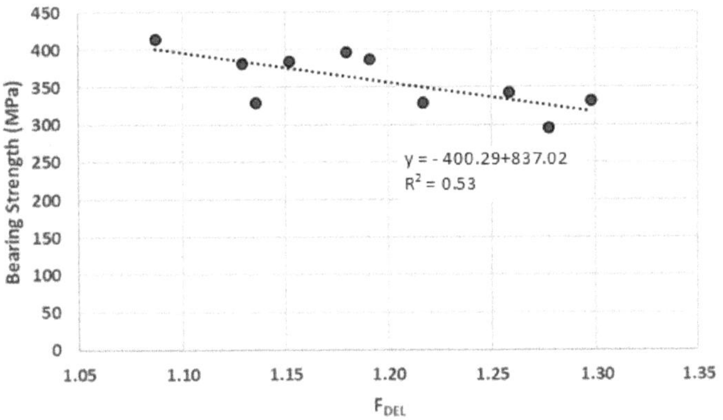

Figure 4. Correlation between Damage and Bearing strength for cross-ply plates.

On the other hand, when carrying out the same analysis for the unidirectional plates, it is possible to conclude that no correlation is possible, as is easy to conclude just by observing the values in Table 4, where the higher bearing strength was obtained with the plates drilled with the HSS drill. In fact, the drilling damage can easily propagate in a direction parallel to fibers (0° direction), leading to the fragile fracture of the coupon, as demonstrated in Figure 5, showing a radiography of a unidirectional plate after drilling and completion of the bearing test. The longitudinal slit is clearly visible. This diverse outcome for cross-ply plates may be because the 90° aligned fibers in these plates will act as barriers to the damage, preventing its propagation. In fact, it is known from Fracture Mechanics that the existence of fibers perpendicular to the direction of crack propagation creates a fiber pulling mechanism, which results in augmented energy dissipation retarding crack expansion. Therefore, cross-ply plates are helpful in preventing the sudden propagation of existing damages in the plate when compared with unidirectional plates.

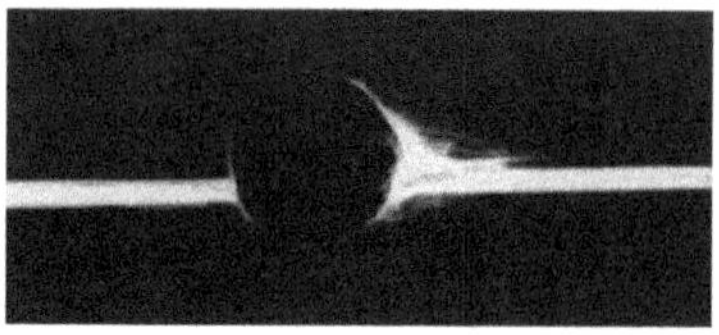

Figure 5. Unidirectional plate after Bearing test—the longitudinal slit is evident.

3.3. Cyclic Tests

The focus of this experimental study was the collection of basic information on the consequences of cyclic loading on bolted connections. This loading is more likely to represent the service conditions of composite plates, simulating the assembling in complex sets and the resultant loading. For that, it is necessary to define how to assess the consequences of cyclic loading. Plates are tested according to ASTM D6873 [38] for a limited number of cycles and then radiographed, making it possible to compare the damage progression along the loading direction, as represented in Figure 6. The hypothesis is the ovality of the delaminated area due to the forced deformation of the hole. Values of loads for cyclic tests are presented in Table 2. This hypothesis was confirmed for cross-ply plates, see Figure 7, showing the sequence of images for a drilled plate after drilling and then after 1000 and 2000 cycles. As stated before, due to operative conditions, each test lasted for approximately 6 h, leading to this selection of the number of cycles. The results of this assessment for the ovality of holes are presented in Table 5. No fracture occurred during these tests for the number of cycles performed.

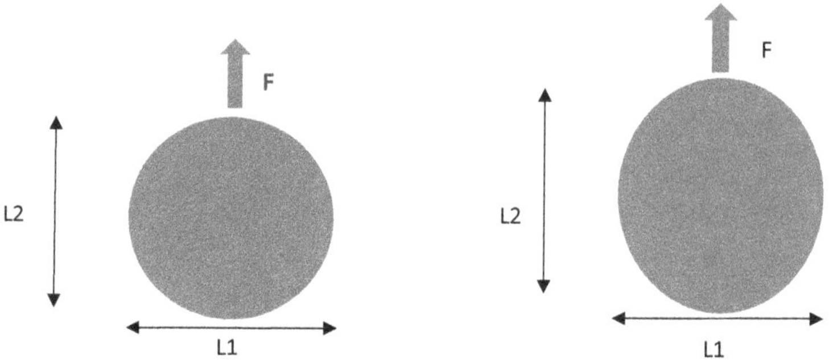

Figure 6. Ovality of drilled holes due to cyclic bearing test—concept.

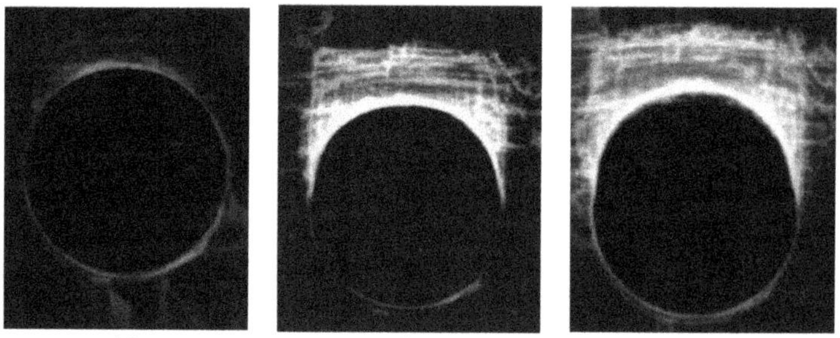

(a) (b) (c)

Figure 7. Radiography of drilled cross-ply plates: (**a**) as drilled; (**b**) after 1000 cycles; (**c**) after 2000 cycles—ovality in the load direction (top) is evident.

Table 5. Ovality of drilled holes after bearing cycles ASTM D-6873 (values in mm).

Drilling Tool	L1	L2/as Drilled	L2/1000 Cycles	L2/2000 Cycles
WC BRAD	6.03	6.04	6.10	6.75
WC TWIST	5.95	6.06	6.14	6.66
HSS TWIST	5.92	5.97	6.17	6.71

For unidirectional plates, ovality was not so relevant; the ovality measured in L2 just differed 0.04 mm from zero to 2000 cycles; however, some plates fractured during the test, just like in the monotonic bearing tests, as in Figure 5, even when loads were reduced to 70% of 60% bearing strength, (Table 2). The ovality for the plates that did not break was not extended, showing that plates with this stacking sequence are not suitable to tolerate deformations. In fact, fracture resulted after the displacement of the movable head of the test machine overcame 0.42 mm. This failure of some plates after a determined number of cycles made it possible to establish a preliminary correlation between load amplitude and the number of cycles to failure, as represented in Figure 8, resembling the well-known Wohler curve for metallic materials. Therefore, it is possible that, with a considerable number of tests, some correlation can be found and a "fatigue curve" can be drawn for these materials adapted to the diversity of the stacking sequences, making it possible to set a fatigue limit.

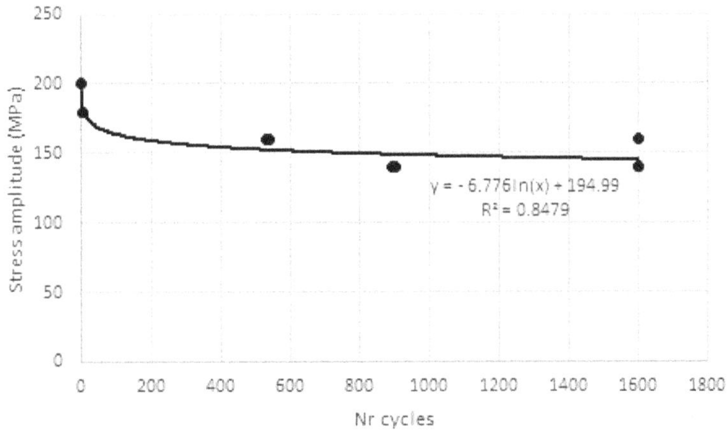

Figure 8. Stress amplitude vs. number of cycles for unidirectional plates.

Finally, Figure 9 represents two examples of typical pin displacement versus the number of cycles observed in two different plates, one with fracture after 530 cycles, where an uprising of displacement is clearly visible prior to test stop, and another that has passed the 1000 cycles test without fracture.

An interesting feature noted in these tests was the higher on average cyclic resistance of holes drilled with HSS drills. This outcome may be the result of higher delamination in scattered directions, which acts as a barrier to fatigue failure progression. Based on the test results and the observation of the damage progression, it is evident that the progression of damage follows the same direction of the applied load, resulting in further cracks and increasing damage extension as the consecutive loads causes the breakage of more fibers. As the damage progresses, the resistant cross-section decreases due to the consecutive failure of fibers and matrix cracking and deformation, even though the matrix material is not mechanically strong but contributes to keep the bulk form of the plate. As the matrix material cracks, the bulk shape of the plate will easily be deformed as the fibers themselves cannot hold the original shape.

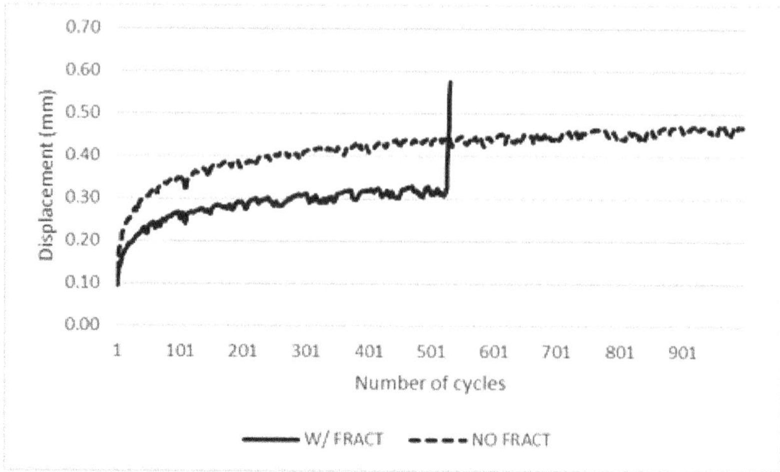

Figure 9. Pin displacement vs. number of cycles for cross-ply plates.

These assumptions must be confirmed with further studies and tests with a larger number of cycles, as is normal in fatigue studies.

3.4. Residual Bearing Strength Tests

The final step of this experimental study was the completion of bearing tests according to ASTM D5961 on the plates that supported 2000 cycles to determine the residual bearing resistance. So, three unidirectional and three cross-ply plates that were cyclic loaded and, consequently, deformed, as detailed in the precedent section, were tested and the results are presented in Table 6, including the bearing ratio, resulting from dividing the bearing strength after cyclic loading by the original average value for each drill condition on tool geometry.

Table 6. Residual bearing strength of plates after cyclic testing.

PLATE	DRILL	DELAMINATION AFTER DRILLING	RESIDUAL BEARING STRENGTH (MPa)	BEARING RATIO
UNIDIRECTIONAL	BRAD	1.12	191	0.96
	TWIST WC	1.12	208	0.97
	TWIST HSS	1.32	202	0.98
CROSS PLY	BRAD	1.18	346	0.91
	TWIST WC	1.12	339	0.95
	TWIST HSS	1.28	314	0.97

For the unidirectional plates, there was no significative difference in the bearing strength due to previous deformation. In these plates, deformation seems to have had little or no effect on bearing strength. The decrease in the maximum bearing load is always small and the global average variation is less than 3%, demonstrating that unidirectional plates are not able to withstand deformation. On the other hand, for hole deformed cross-ply plates, the bearing strength decreased by around 9% if we consider Brad drilled plates, and by 7.5% on the global average of all valid test results. It is interesting to note that the wider the damage, the less the bearing strength decreases. In [28], the authors concluded a slight increase in bearing strength due to blunting. Notice that in [28], the number of cycles was 10^6, much more than the number of cycles carried out in this study. So, it is reasonable to say that the blunting of the hole surrounding area can contribute to enhance fatigue resistance by hole distortion, which can be confirmed visually. All of these ratios need to be confirmed by performing a larger number of tests with a higher number of cycles. The rise in mechanical resistance to subsequent deformation due to hole distortion is a recognized reality in bearing strength tests. In ASTM D5961 standard [46], it is noted that it is possible to prevent masking of the true failure mode by large-scale hole distortion, as evidenced and confirmed within this study.

4. Conclusions

Here, we present an exploratory study on the conditions of damage propagation for fatigue testing on drilled composite plates in carbon/epoxy. The motivation was to gather a set of results for the future planning of long-term fatigue testing, which is time demanding and requires specialized equipment. For this purpose, two batches of carbon/epoxy plates, one with a unidirectional and the other with a cross-ply stacking sequence, were prepared for the above-described experimental sequence, including drilling, which is necessary for assembling purposes, and damage assessment by image processing, using existing damage criteria and low frequency cyclic testing, following ASTM D6873 procedure. From the work performed, some conclusions were made as follows:

- By using enhanced radiography, both for delamination measurement and ovality progression, with the help of MatLab® Image Processing tools, the delamination assessment procedure can assist in measuring damage progression;
- As the drilling process causes more damage around the hole, the bearing strength of the drilled plate decreases, evidencing the importance of proper cutting parameters to enhance reliability in service;
- As predicted, fatigue loading causes ovality of machined holes and decreases the bearing load capacity of machined holes after a certain number of cycles;
- The response to cyclic tests is different for the two stacking sequences of this work, showing that unidirectional plates are not suitable for load bearing or cyclic fatigue bearing loads. For cross-ply plates, it is possible to identify an asymptote of the fatigue vs. number of cycles curve.

Additionally, further studies are recommended based on these primary conclusions.

Cyclic tests need be performed with a larger number of cycles, from 20,000 to 2×10^6 at 5 Hz frequency, with an adequate setup for image recording and interrupted at defined moments for radiography and ovality assessment;

An extended number of tests to study the possible effect of machining parameters on fatigue resistance, as well as stacking sequence effects, should be performed;

Another possible path of exploration is the use of hybrid composites with carbon and glass fibers as reinforcement.

Author Contributions: Conceptualization, L.M.P.D. and J.L.E.; methodology, N.C.L.; software, J.E.M.; validation, N.C.L. and S.C.F.F.; formal analysis, S.C.F.F.; investigation, N.C.L.; resources, L.M.P.D. and J.L.E.; data curation, J.E.M.; writing—original draft preparation, S.C.F.F.; writing—review and editing, J.L.E.; visualization, J.E.M.; supervision, L.M.P.D.; project administration, J.L.E.; funding acquisition, L.M.P.D. All authors have read and agreed to the published version of the manuscript.

Funding: This work was granted by the Portuguese FCT in the scope of the project UID/EMS/0615/2019.

Informed Consent Statement: Not applicable.

Data Availability Statement: Research data is unavailable.

Conflicts of Interest: The authors declare no conflict of interest.

References

1. Krishnaraj, V.; Zitoune, R.; Davim, J.P. *Drilling of Polymer-Matrix Composites*; Briefs in Manufacturing and Surface Engineering; Springer: Berlin/Heidelberg, Germany, 2013; ISBN 978-3-642-38344-1.
2. Soutis, C. Fibre reinforced composites in aircraft construction. *Prog. Aerosp. Sci.* **2005**, *41*, 143–151. [CrossRef]

3. Bhagwan, D.; Agarwal, B.D.; Broutman, L.J.; Chandrashekhara, K. *Analysis and Performance of Fiber Composites*, 4th ed.; Wiley: Hoboken, NJ, USA, 2017; ISBN 978-1-119-38998-9.
4. Ahmad, H.; Markina, A.A.; Porotnikov, M.V.; Ahmad, F. A review of carbon fiber materials in automotive industry. *IOP Conf. Series Mater. Sci. Eng.* **2020**, *971*, 0320112020. [CrossRef]
5. Sreejith, M.; Rajeev, R. *Fiber Reinforced Composites for Aerospace and Sports Applications*; Series in Composites Science and Engineering; Woodhead Publishing: Cambridge, UK, 2021; pp. 821–859.
6. Ghabezi, P.; Khoran, M. Optimization of Drilling Parameters in Composite Sandwich Structures (PVC Core). *Indian J. Sci. Res.* **2014**, *2*, 173–179.
7. Ghabezi, P.; Farahani, M.; Shahmirzaloo, A.; Ghorbani, H.; Harrison, N.M. Defect evaluation of the honeycomb structures formed during the drilling process. *Int. J. Damage Mech.* **2020**, *29*, 454–466. [CrossRef]
8. Yasar, N.K.; Korkmaz, M.E.; Günay, M. Investigation on hole quality of cutting conditions in drilling of CFRP composite. *MATEC Web Conf.* **2017**, *112*, 01013. [CrossRef]
9. Rajkumar, D.; Ranjithkumar, P.; Jenarthanan, M.P.; Sathiya Narayanan, C. Experimental investigation and analysis of factors influencing delamination and thrust force during drilling of carbon-fibre reinforced polymer composites. *Pigment. Resin Technol* **2017**, *46*, 507–524.
10. Xu, J.; Li, L.; Geier, N.; Davim, J.P.; Chen, M. Experimental study of drilling behaviors and damage issues for woven GFRP composites using special drills. *J. Mater. Res. Technol.* **2022**, *21*, 1256–1273. [CrossRef]
11. Geng, D.; Liu, Y.; Shao, Z.; Lu, Z.; Cai, J.; Li, X.; Jiang, X.; Zhang, D. Delamination formation, evaluation and suppression during drilling of composite laminates: A review. *Compos. Struct.* **2019**, *216*, 168–186. [CrossRef]
12. Durão, L.M.P. Machining of Fiber Reinforced Composites. In *Fiber Reinforced Composites*; Cheng, Q., Ed.; Nova Science Publishers: New York, NY, USA, 2012; pp. 387–438. ISBN 9781614703037.
13. Abrate, S. *Composites Engineering Handbook*; Mallick, P.K., Ed.; Marcel Dekker: New York, NY, USA, 1997; pp. 777–809.
14. Wern, C.W.; Ramulu, M.; Shukla, A. Investigation of stresses in the orthogonal cutting of fiber-reinforced plastics. *Exp. Mech.* **1996**, *36*, 33–41. [CrossRef]
15. Hocheng, H.; Tsao, C.C. The path towards delamination free drilling of composite materials. *J. Mater. Process. Technol.* **2005**, *167*, 251–264. [CrossRef]
16. Won, M.S.; Dharan, C.K.H. Chisel Edge and Pilot Hole Effects in Drilling Composite Laminates. *J. Manuf. Sci. Eng.* **2002**, *124*, 242–247. [CrossRef]
17. Hocheng, H.; Dharan, C.K.H. Delamination During Drilling in Composite Laminates. *J. Eng. Ind.* **1990**, *112*, 236–239. [CrossRef]
18. Hocheng, H.; Tsao, C.C. Comprehensive analysis of delamination in drilling of composite materials with various drill bits. *J. Mater. Process. Technol.* **2003**, *140*, 335–339. [CrossRef]
19. Durão, L.M.P.; Magalhães, A.; Marques, A.T.; Baptista, A.; Figueiredo, M. Drilling of Fibre Reinforced Plastic Laminates. *Mater. Sci. Forum* **2008**, *587–588*, 706–710.
20. Durão, L.M.P.; Gonçalves, D.J.; Tavares, J.M.R.; de Albuquerque, V.H.C.; Vieira, A.A.; Marques, A.T. Drilling tool geometry evaluation for reinforced composite laminates. *Compos. Struct.* **2010**, *92*, 1545–1550. [CrossRef]
21. Marques, A.T.; Durão, L.M.; Magalhães, A.G.; Silva, J.F.; Tavares, J.M.R.S. Delamination analysis of carbon fibre reinforced laminates: Evaluation of a special step drill. *Compos. Sci. Technol.* **2009**, *69*, 2376–2382. [CrossRef]
22. Tsao, C.C. Effect of pilot hole on thrust force by saw drill. *Int. J. Mach. Tools Manuf.* **2007**, *47*, 2172–2176. [CrossRef]
23. Persson, E.; Ingvar Eriksson, I.; Zackrisson, L. Effects of hole machining defects on strength and fatigue life of composite laminates. *Compos. Part A Appl. Sci. Manuf.* **1997**, *28*, 141–151. [CrossRef]
24. Talreja, R. Damage mechanics and fatigue life assessment of composite materials. *Int. J. Damage Mech.* **1999**, *8*, 339–354. [CrossRef]
25. Talreja, R. Fatigue of Composite Materials. In *Modern Trends in Composite Laminates Mechanics*; Altenbach, H., Becker, W., Eds.; Springer: Wien, Austria, 2003; pp. 281–294.
26. Nixon-Pearson, O.J.; Hallett, S.R.; Withers, P.J.; Rouse, J. Damage development in open-hole composite specimens in fatigue. Part 1: Experimental investigation. *Compos. Struct.* **2013**, *106*, 882–889. [CrossRef]
27. Nixon-Pearson, O.J.; Hallett, S.R.; Harper, P.W.; Kawashita, L.F. Damage development in open-hole composite specimens in fatigue. Part 2: Numerical modelling. *Compos. Struct. Compos. Struct.* **2013**, *106*, 890–898. [CrossRef]
28. Nixon-Pearson, O.J.; Hallett, S.R. An investigation into the damage development and residual strengths of open-hole specimens in fatigue. *Compos. Part A* **2015**, *69*, 266–278. [CrossRef]
29. Hochard, C.; Miot, S.; Thollon, Y. Fatigue of laminated composite structures with stress concentrations. *Compos. Part B* **2014**, *65*, 11–16. [CrossRef]
30. Muc, A.; Romanowicz, P. Effect of notch on static and fatigue performance of multilayered composite structures under tensile loads. *Compos. Struct.* **2017**, *178*, 27–36. [CrossRef]
31. Yenigun, B.; Kilickap, E. Influence of hole quality on fatigue life of drilled CFRP with the different ply orientation angle. *J. Braz. Soc. Mech. Sci. Eng.* **2021**, *43*, 1–14. [CrossRef]
32. Van Der Sypt, P.; Chérif, M.; Bois, C. Analysis of the fatigue behaviour of laminated composite holes subjected to pin-bearing loads. *Int. J. Fatigue* **2017**, *103*, 86–98. [CrossRef]
33. Guo, R.; Xian, G.; Li, C.; Hong, B. Effect of fiber hybrid mode on the tension–tension fatigue performance for the pultruded carbon/glass fiber reinforced polymer composite rod. *Eng. Fract. Mech.* **2022**, *260*, 108208. [CrossRef]

34. Camanho, P.P.; Erçin, G.H.; Catalanotti, G.; Mahdi, S.; Linde, P. A finite fracture mechanics model for the prediction of the open-hole strength of composite laminates. *Compos. Part A* **2012**, *43*, 1219–1225. [CrossRef]
35. Bao, H.; Liu, G. Progressive failure analysis on scaled open-hole tensile composite laminates. *Compos. Struct.* **2016**, *150*, 173–180. [CrossRef]
36. Hubert Debski, H.; Samborski, S.; Rozylo, P.; Wysmulski, P. Stability and Load-Carrying Capacity of Thin-Walled FRP Composite Z-Profiles under Eccentric Compression. *Materials* **2020**, *13*, 2956. [CrossRef]
37. Wysmulski, P. Non-linear analysis of the postbuckling behaviour of eccentrically compressed composite channel-section columns. *Compos. Struct.* **2023**, *305*, 116446. [CrossRef]
38. *ASTM D6873-19*; Standard Practice for Bearing Fatigue Response of Polymer Matrix Composite Laminates. ASTM International: West Conshohocken, PA, USA, 2019.
39. Durão, L.M.P.; Gonçalves, D.J.S.; Tavares, J.M.R.S.; de Albuquerque, V.H.C.; Marques, A.T. Comparative analysis of drills for composite laminates. *J. Compos. Mater.* **2011**, *46*, 1649–1659. [CrossRef]
40. Luís Miguel, P.; Durão, L.M.P.; Tavares, J.M.R.S.; de Albuquerque, V.H.C.; Marques, J.F.S.; Andrade, O.N.G. Drilling Damage in Composite Material. *Materials* **2014**, *7*, 3802–3819.
41. Composite Materials Italy. Available online: https://www.composite-materials.it/pagina.php?cod=1 (accessed on 24 September 2022).
42. De Albuquerque, V.H.C.; Tavares, J.M.R.S.; Durão, L.M.P. Evaluation of delamination damages on composite plates from radiographic image processing using an artificial neural network. *J. Compos. Mater.* **2010**, *44*, 1139–1159. [CrossRef]
43. Durão, L.M.P.; Magalhães, A.G.; Tavares, J.M.R.S.; Marques, A.T. Analyzing objects in images for estimating the delamination influence on load carrying capacity of composite laminates. *Electron. Lett. Comput. Vis. Image Anal.* **2008**, *7*, 11–21.
44. Bajpai, P.K.; Debnath, K.; Singh, I. Hole making in natural fiber-reinforced polylactic acid laminates: An experimental investigation. *J. Thermoplast. Compos. Mater.* **2017**, *30*, 30–46. [CrossRef]
45. Durão, L.M.; Panzera, T.H.; Scarpa, F.; Sérgio Filho, L.M.; Oliveira, P.R. Damage assessment of fibre reinforced laminates. *Compos. Struct.* **2015**, *133*, 939–946. [CrossRef]
46. *ASTM D5961-17*; Standard Test Method for Bearing Response of Polymer Matrix Compo-Site Laminates. ASTM International: West Conshohocken, PA, USA, 2021.

materials

MDPI

Article

Micro-Scale Numerical Simulation of Fatigue Failure for CFRP Subjected to Multiple-Amplitude Cyclic Loadings Based on Entropy Damage Criterion

Huachao Deng [1], Keitaro Toda [1], Mio Sato [2] and Jun Koyanagi [1,*]

[1] Department of Materials Science and Technology, Tokyo University of Science, Tokyo 125-8585, Japan
[2] Japan Aerospace Exploration Agency, Osawa, Tokyo 181-0015, Japan
* Correspondence: koyanagi@rs.tus.ac.jp

Abstract: Fatigue failure of carbon fiber-reinforced plastics (CFRPs) under cyclic loadings has attracted the attention of researchers recently. In this study, the entropy-based failure criterion is proposed to investigate the fatigue lifetime of unidirectional CFRPs subjected to multiple-amplitude cyclic loadings. Due to the heterogeneity of CFRPs, a micro-finite element model considering matrix resin and fibers independently is developed, and the entropy-based damage criterion is implemented into a user-subroutine of Abaqus to model the progressive damage of matrix resin. The fatigue lifetime of CFRPs under typical loading sequences consisting of two stages, such as varying from low to high (L-H) or from high to low (H-L) loading sequence, is estimated with the proposed failure criterion. Numerical results show that the initial damage occurs near the area between two fibers, and a transverse crack propagates progressively under the cyclic loading. The difference in predicted lifetime to final failure in L-H and H-L stress levels is 6.3%. Thus, the effect of loading sequence on the fatigue lifetime can be revealed via the proposed entropy-based damage criterion. Comparisons with the conventional linear cumulative damage (LCD) and kinetic crack growth (KCG) theories are also conducted to demonstrate the validity of the proposed method. The entropy-based failure criterion is a promising method to predict the residual strength and fatigue lifetime of CFRP components.

Keywords: matrix resin failure; entropy-based strength degradation; multiple-amplitude cyclic loadings; finite element simulation

Citation: Deng, H.; Toda, K.; Sato, M.; Koyanagi, J. Micro-Scale Numerical Simulation of Fatigue Failure for CFRP Subjected to Multiple-Amplitude Cyclic Loadings Based on Entropy Damage Criterion. *Materials* **2023**, *16*, 6120. https://doi.org/10.3390/ma16186120

Academic Editor: Aniello Riccio

Received: 15 August 2023
Revised: 1 September 2023
Accepted: 4 September 2023
Published: 7 September 2023

1. Introduction

Due to their excellent corrosion resistance, specific strength and modulus, CFRPs are being widely utilized in the aerospace [1,2], automotive [3,4], construction industries [5,6], lightweight structures [7,8] and military fields [9,10]. Differing from conventional materials, CFRPs usually consist of a polymer matrix and fibers [11], in which the polymer matrix usually exhibits visco-elastic behavior [12]. During service periods, CFRPs are subjected to arbitrary and multiple-amplitude loadings, and the failure mechanisms, such as transverse cracks [13–16] and interface delamination [17], are also time- and temperature-dependent [18]. It should be noted that the loading sequences are usually randomly applied on the CFRP components [19–21]. Accurate prediction of fatigue lifetimes is crucial for the integrity of CFRP components [22,23]. However, investigations into the fatigue lifetime of CFRPs under random loadings are still limited. Thus, the development of effective methodologies to estimate the fatigue lifetimes of CFRP components under random loadings [24,25] is of great significance.

In early works, the investigations into lifetime estimations of CFRPs were mainly conducted using the experimental method. Based on the experimental results [26], it was found that fiber-reinforced plastics suffered from fiber breakage, transverse cracks and interface delamination, in which the transverse cracks accounted for a large portion of the total period until final fatigue failure. Ogihara et al. [27] conducted a constant amplitude fatigue test of quasi-isotropic CFRP laminates, and experimental results revealed that the load level and the number of cycles were two significant factors affecting transverse crack accumulation. Li et al. [28] investigated the effects of elevated temperature and hydraulic pressure on the long-term properties of a carbon/glass hybrid rod. In addition to the stress level and number of loading cycles, it was also known that load frequency was a key factor affecting the fatigue performance of polymer–matrix composites. Miyano et al. [29] studied the fatigue strength of CFRP laminates and developed a prediction method to address the effects of load frequency and stress ratio based on certain hypotheses. It is worth noting that under the given loading cycles, a lower loading frequency would result in greater damage than a higher one [30,31]. Additionally, the applications of CFRPs for lightweight structures [32–34] are also notable. Chen et al. [35] tested the mechanical performance and investigated the progressive failure characteristics of CFRP/aluminum joints for lightweight applications. Zhang and An [36] proposed a topology optimization of composite structures for additive manufacturing. To analyze the failure behaviors of CFRPs for lightweight vehicle applications, Sun et al. [37] developed an integrated computational materials–engineering framework at the same time.

Recently, micro-scale modeling was also proposed to investigate the failure mechanisms of CFRPs, in which both the polymer matrix and fibers were considered. Asp et al. [38] presented a scheme of critical dilatational deformation in the first quadrant of the bi-axial failure envelope for polymers, which physically suggested that cavitation or crazing occurs in polymer materials. Based on the work of Asp et al. [38], a strain-invariant failure theory and element-failure method were also developed [39,40]. Canal et al. [41] implemented a periodic unit-cell simulation of matrix failure with an elasto-visco-elastic constitutive law. However, the above-mentioned literature only addresses static failure. To address the time-dependence of the matrix, Schapery [12] proposed methods for evaluating the properties of polymeric materials with some specific constitutive equations derived from thermodynamic principles. McCartney [42,43] derived a theory of crack propagation in a linear visco-elastic material, based on an energy balance fracture criterion. Based on the linear cumulative damage law (Miner's law), Christensen [44,45] developed the kinetic crack growth theory to estimate the lifetime of polymeric materials. Additionally, Koyanagi et al. [46] developed an elasto-visco-plastic constitutive equation using continuum damage mechanics to simulate the strain-rate dependent transition of the transverse tensile failure mode in fiber-reinforced composites. Numerical results showed that in the case of high strain rates, interface failure was the dominant failure mode, while matrix failure was dominant at low strain rates. Nevertheless, the existing methods were not suitable for estimations of the fatigue lifetime and residual strength of CFRPs subjected to randomly loading sequences.

According to thermodynamics, when solid materials are subjected to thermomechanical loadings, the entropy generation [47] is always non-negative. Final material failure occurs when the entropy inside the material reaches a critical value. Thus, the entropy generation of solid materials under cyclic loading can be used to reveal the degradation of properties and estimate the fatigue lifetime and residual strength of structures. It is worth noting that Naderi et al. [48] have conducted a series of fatigue tests to examine the validity of the damage criterion based on the entropy. Results showed that the cumulative entropy generation was constant at the time of final failure and was independent of geometrical shape, stress level and frequency. At the instance of final failure, the total entropy generation was defined as the fatigue fracture entropy.

To the best knowledge of authors, the applications of an entropy-based failure criterion to estimate the residual strength of CFRPs are still limited [49,50]. Koyanagi et al. [51–53] recently developed a computational framework with an entropy-based failure criterion to study the failure mechanism of a visco-elastic matrix and CFRP cross-ply laminates. It was found that the entropy-based failure criterion can reproduce the effect of load frequency on fatigue failure.

However, the estimation of fatigue lifetimes of CFRPs under randomly loading sequences using the entropy-based failure criterion has not been reported. In this study, the entropy-based failure criterion is proposed to estimate the fatigue lifetime of CFRPs subjected to multi-amplitude cyclic loadings. The organization of the paper is as follows: the entropy-based failure criterion is presented in Section 2; numerical simulations and discussions are presented in Section 3 and the conclusions are drawn in Section 4.

2. Numerical Method

2.1. Visco-Elastic Constitutive Law

Due to its complexity, the deformation and fracture behaviors of CFRPs are found to be much more complicated than homogeneous materials. In particular, the matrix of CFRPs consists of polymer resin whose properties exhibit high time-dependency. In this study, the visco-elastic constitutive law, based on Maxwell elements, is utilized to characterize the viscosity of the matrix resin, where the number of Maxwell elements is 15 [52], as shown in Figure 1.

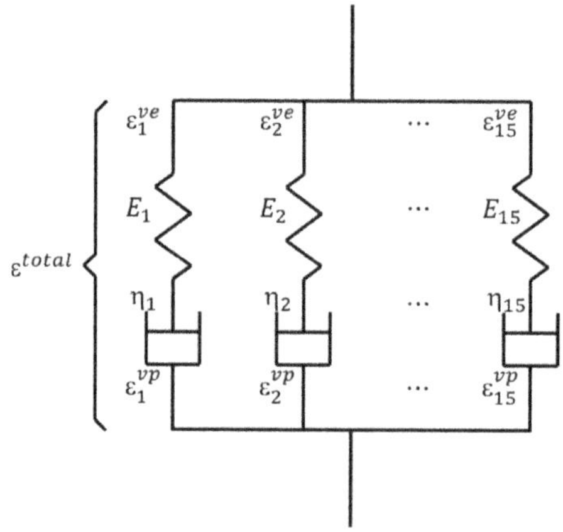

Figure 1. Visco-elastic model with 15 Maxwell elements.

In this constitutive law, the total strain ε consists of a visco-elastic strain ε^{ve} and a visco-plastic strain ε^{vp} [12]:

$$\varepsilon = \varepsilon^{ve} + \varepsilon^{vp} \tag{1}$$

In Equation (1), the visco-plastic strain ε^{vp} depends on viscosity and is calculated as

$$\varepsilon^{vp} = \int_0^t \mathbf{H}^{-1}\sigma dt \tag{2}$$

where

$$\mathbf{H} = \frac{\eta^{vp}}{(1+v)(1-2v)}\mathbf{M} \tag{3}$$

$$\mathbf{M} = \begin{bmatrix} 1-v & v & v & 0 & 0 & 0 \\ & 1-v & v & 0 & 0 & 0 \\ & & 1-v & 0 & 0 & 0 \\ & & & \frac{1}{2}-v & 0 & 0 \\ & Symmetry & & & \frac{1}{2}-v & 0 \\ & & & & & \frac{1}{2}-v \end{bmatrix} \tag{4}$$

$$\eta^{vp} = \eta_0 \times \frac{1 + e^{\beta(\varepsilon_{eqv}^{vp}/\sigma_{eqv})^{\chi}}}{1 + e^{\alpha(\sigma_{eqv} - \sigma_{vp0})}} \tag{5}$$

In Equations (3)–(5), v is Poisson's ratio, η_0 is the initial viscosity, α, β and χ are coefficients, $\sigma_{eqv} = \sqrt{\frac{3}{2}\mathbf{s} : \mathbf{s}}$ and $\varepsilon_{eqv}^{vp} = \sqrt{\frac{2}{3}\boldsymbol{\varepsilon}^{vp} : \boldsymbol{\varepsilon}^{vp}}$ are equivalent stress and visco-plastic strain and σ_{vp0} is specific stress.

One visco-elastic constitutive law considering the damage D is expressed as [54]

$$\boldsymbol{\sigma}(t) = (1 - D)\int_0^t \mathbf{E}^r(t - t')g\dot{\boldsymbol{\varepsilon}}^{ve}\mathrm{d}t' \tag{6}$$

where the relaxation modulus $\mathbf{E}^r$ is

$$\mathbf{E}^r(t) = \sum_{n=1}^{15} E_{ijkl}^n e^{-tE^n/\eta^n} \tag{7}$$

And the nonlinear coefficient is

$$g = \frac{1}{1 + \alpha\left(\frac{\sigma_{eqv}}{\sigma_0}\right)^m} \tag{8}$$

2.2. Entropy-Based Failure Criterion

Based on the second law of thermodynamics (the Clausius–Duhem inequality) [48], the irreversible entropy generation in the material is calculated as

$$s = \int_0^t \left(\frac{1}{T}\boldsymbol{\sigma} : \dot{\boldsymbol{\varepsilon}}^{vp} - \frac{\boldsymbol{A}\cdot\boldsymbol{V}}{T} - \boldsymbol{Q}\cdot\frac{\nabla T}{T^2}\right)\mathrm{d}t \tag{9}$$

where the generalized thermodynamic force vector $\boldsymbol{A}$ is conjugate with the generalized thermodynamic internal flow vector $\boldsymbol{V}$ and $\boldsymbol{Q}$ is the heat flux vector. Practically, entropy generation owing to plastic deformation is dominant, and the last two terms in Equation (9) are negligible. Thus, Equation (9) reduces to [48]

$$s = \int_0^t \frac{1}{T}\boldsymbol{\sigma} : \dot{\boldsymbol{\varepsilon}}^{vp}\mathrm{d}t \tag{10}$$

Using Equation (10), the final fracture entropy s^f per unit volume of materials can be calculated using the constitutive law or experimental method. When the material is subjected to cyclic loading, the damage variable D, associated with entropy generation s, is introduced to characterize the progressive degradation of material and expressed as [52]

$$D = \frac{s}{s^f}D_{cr} \tag{11}$$

where D_{cr} is the user-defined critical damage. In this study, the final fracture entropy s^f is assumed to be independent of loading conditions, such as stress level and frequency.

2.3. Implementation of the Proposed Method

The proposed entropy-based failure criterion is implemented into a user subroutine to define a material's mechanical behavior (UMAT) in Abaqus (2020, Dassault Système), and illustrated via a flowchart in Figure 2. With the user subroutine interface, the increments of strain, stress and solution-dependent variables, such as the Maxwell element's stress $\boldsymbol{\sigma}_n^{old}$ and damage D, are passed into UMAT. Based on these, σ_{eqv} and ε_{eqv}^{vp} are firstly computed to determine viscosity η^{vp}, nonlinear coefficient g and matrices $\mathbf{D}$ and $\mathbf{H}$. Additionally, the visco-plastic strain increment $\Delta\boldsymbol{\varepsilon}^{vp}$ is computed to update the increment of visco-elastic strain $\Delta\boldsymbol{\varepsilon}^{ve}$ and temporal stress increment $\Delta\boldsymbol{\sigma}^{temp}$. Then, the temporal spring and dashpot elongations are calculated to update the stress increment of the Maxwell elements $\Delta\sigma_n$. Finally, the stress at the integration point $\boldsymbol{\sigma}$, dissipated energy increment ΔE and entropy increment Δs are computed to update the damage D and damage increment ΔD.

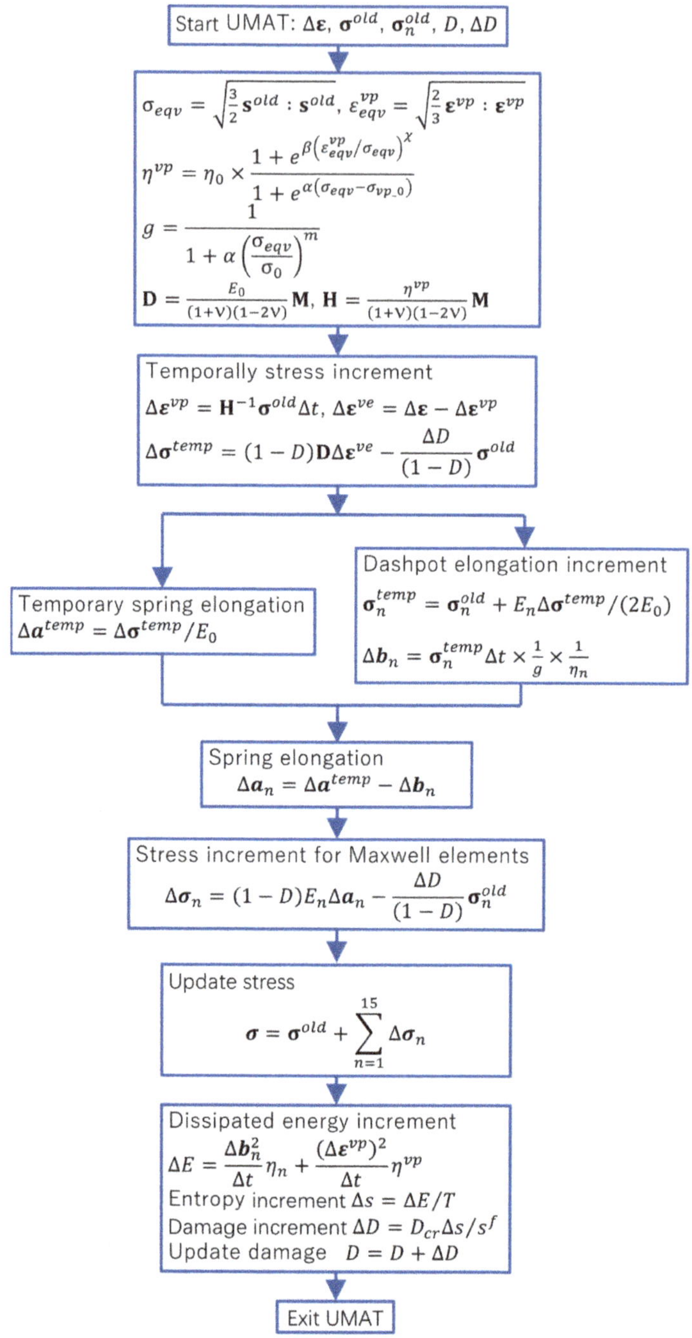

Figure 2. Illustration of updating stress via entropy-based failure criterion.

3. Numerical Results and Discussions

In this section, the entropy-based failure criterion is applied to estimate the lifetime of CFRPs under a multiple-amplitude loading sequence. As shown in Figure 3, a three-dimensional (3D) unit cell with dimensions of 39 μm × 39 μm × 0.3 μm [52,55] is utilized to conduct the numerical simulation, in which the diameter of the fibers is 6 μm, the number of fibers is 37 and the volume fraction is 56%. The discretization scheme of this model consists of 36,718 nodes and 18,122 C3D8 elements. The carbon fiber is assumed to be an orthotropic elastic material and its properties are listed in Table 1. The matrix resin is modeled as a visco-elastic medium whose constitutive law is implemented using a user-subroutine UMAT of Abaqus, and its properties are listed in Table 2. A larger α_d is chosen to accelerate the evolution of damage and to reduce the simulation time. The displacement boundary condition and external force are employed via the method in the references [52,55–57]. In the case of multiple-amplitude cyclic loadings, the loading sequence only consists of two stages, illustrated in Figure 4, both of which are considered in conducting the numerical simulation, in which the lower cyclic loading sequence is expressed as $f_{low} = f_0\left[\frac{1}{2}sin\left(\pi t + \frac{3}{2}\pi\right) + \frac{1}{2}\right]$, and the higher sequence is $f_{high} = 2f_{low}$, where $f_0 = 1.17 \times 10^{-4}$.

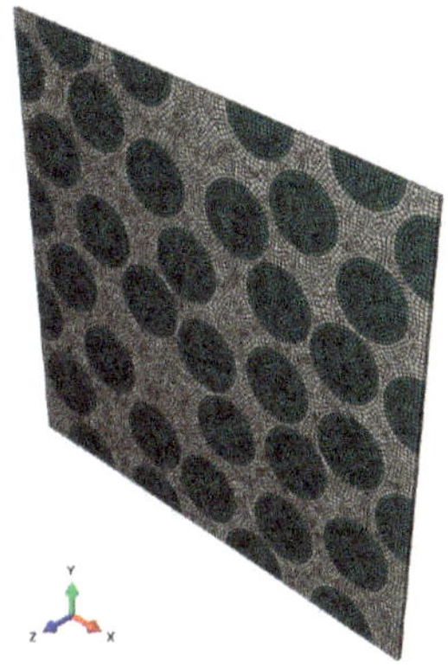

Figure 3. A 3D unit cell model with 37 fibers: green mesh denotes fiber and white is resin.

Table 1. Material properties of fiber.

Properties	E_{11}, E_{22}(GPa)	E_{33}(GPa)	v_{13}, v_{23}	v_{12}
Value	294	15	0.02	0.3

Table 2. Material properties of matrix resin.

n	E_n(MPa)	η_n(MPa·s)	Elasticity	
1	284	4.5×10^2	E_0(MPa)	4260
2	284	3.3×10^3	v	0.3
3	284	1.2×10^5	Nonlinearity	
4	284	1.9×10^6	σ_0(MPa)	70
5	284	1.8×10^7	α	2
6	284	1.4×10^8	m	7
7	284	8.5×10^8	Visco-plastic strain	
8	284	5.0×10^9	η_0(MPa·s)	1.0×10^{23}
9	284	3.0×10^{10}	σ_{vp_0}(MPa)	0
10	284	1.9×10^{11}	α_{vp}	0
11	284	1.4×10^{16}	β_{vp}	0
12	284	1.3×10^{19}	χ	0
13	284	2.1×10^{22}		
14	284	1.3×10^{26}	Damage variables	
15	284	2.5×10^{29}	α_d	4

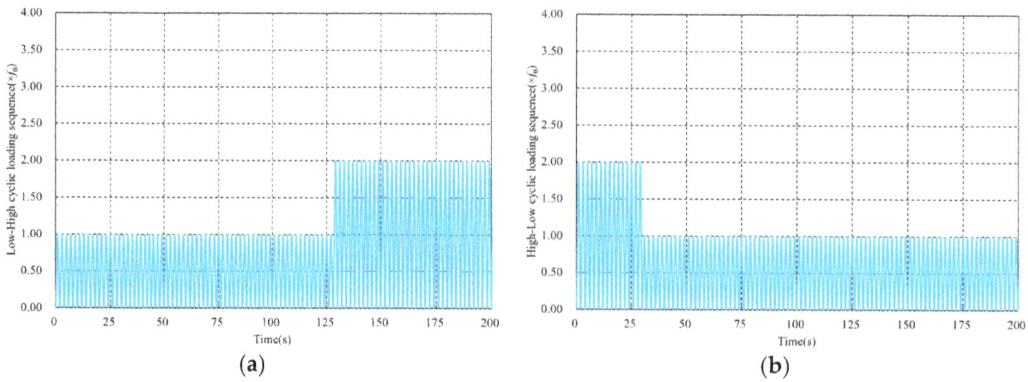

Figure 4. Illustrations of multiple-amplitude cyclic loading sequences. (**a**) Low-high cyclic loading sequence. (**b**) High-low cyclic loading sequence.

3.1. *Fatigue Lifetime under Multiple-Amplitude Loading*

Figure 5 shows the damage evolution of CFRPs under cyclic L-H loading sequences, in which the number of cyclic loadings of the lower level is 64. It can be found that the damage generation initially occurs around the region where two fibers are close to each other. This can be explained by the fact that the deformation variables, such as strain and displacement fields, near this region exhibit significant discontinuity due to the material mismatch between the matrix resin and fibers. In this study, the matrix resin is assumed to be fully failed when the damage variable increases to 0.25 [46], after which the stress will be a very low value, and the initial micro-cracks will be formed between the fiber and the matrix. Additionally, the damage evolutions of the H-L case, where the number of first state sequences is 15, are shown in Figure 6. Analogous to the observations of the L-H loading sequence, the damage accumulations can also be found near the critical region. This phenomenon of damage evolution can also be found in the works of Sato et al. [52], in which only monotonic loading was addressed. Differing from the results of Sato et al. [52], the evolution of accumulated damage caused by the generation of entropy is originally simulated in the current study. As the number of cyclic loadings increases, the damage accumulation is more obvious due to the entropy generation caused by the visco-elastic property of the matrix resin. Thus, the proposed entropy-based failure criterion can reveal the fatigue damage evolution qualitatively.

Figures 7 and 8 show the variations in stress σ_{22} and damage variable D in the earliest element to completely fail. From the initial stage, the dashpots of the Maxwell element move individually due to the cyclic loading, and the dissipated energy gradually accumulates at the same time. The entropy generation can be determined by dividing the dissipated energy by temperature to update the damage variable and degrade the material properties. The higher the loading sequence, the faster the entropy or damage variable increases. Thus, the effect of the loading sequence on the fatigue lifetime of CFRPs can be revealed. In the L-H loading sequence, the damage variable increases to 0.1316 after 64 lower cyclic loadings, after which it reaches 0.25 with 15 higher cyclic loadings. Thus, the total lifetime to full failure can be determined as 64 + 15 = 79. However, in the H-L case, i.e., when the order of lower and higher loading sequences are exchanged, where the number of cyclic loadings in the higher level is taken as 15 (that is, the same as the value in the case of L-H), the damage variable reaches 0.1380 after the first higher level, and 0.25 after 59 lower cyclic loadings. Thus, the predicted total lifetime to final failure in the H-L case is 15 + 59 = 74.

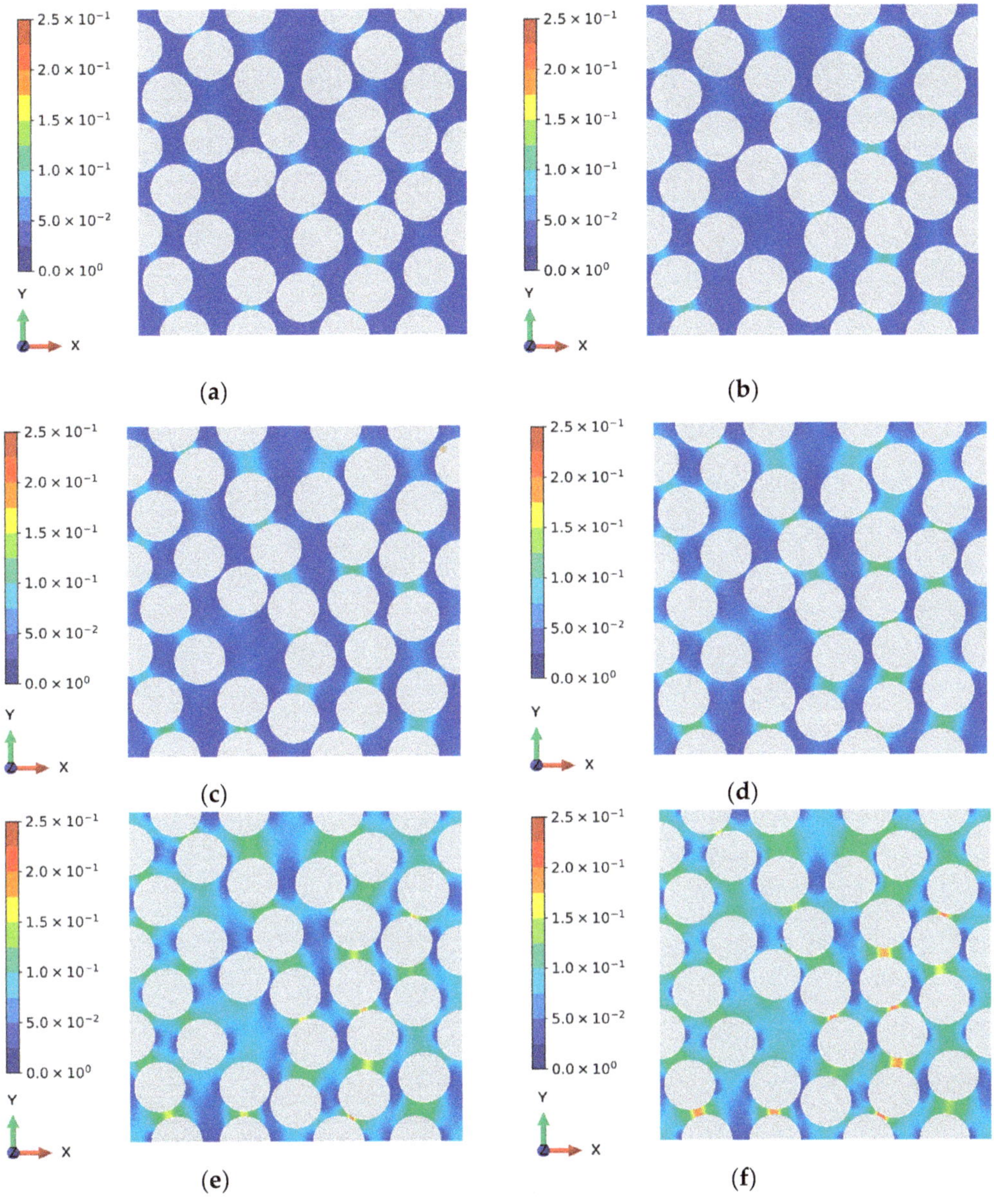

Figure 5. Evolution of damage in CFRPs under low-high cyclic loading sequence. (Palette represents the damage). (**a**) Low stress level: 30 cyclic loadings. (**b**) Low stress level: 40 cyclic loadings. (**c**) Low stress level: 50 cyclic loadings. (**d**) Low stress level: 60 cyclic loadings. (**e**) High stress level: 10 cyclic loadings. (**f**) High stress level: 15 cyclic loadings.

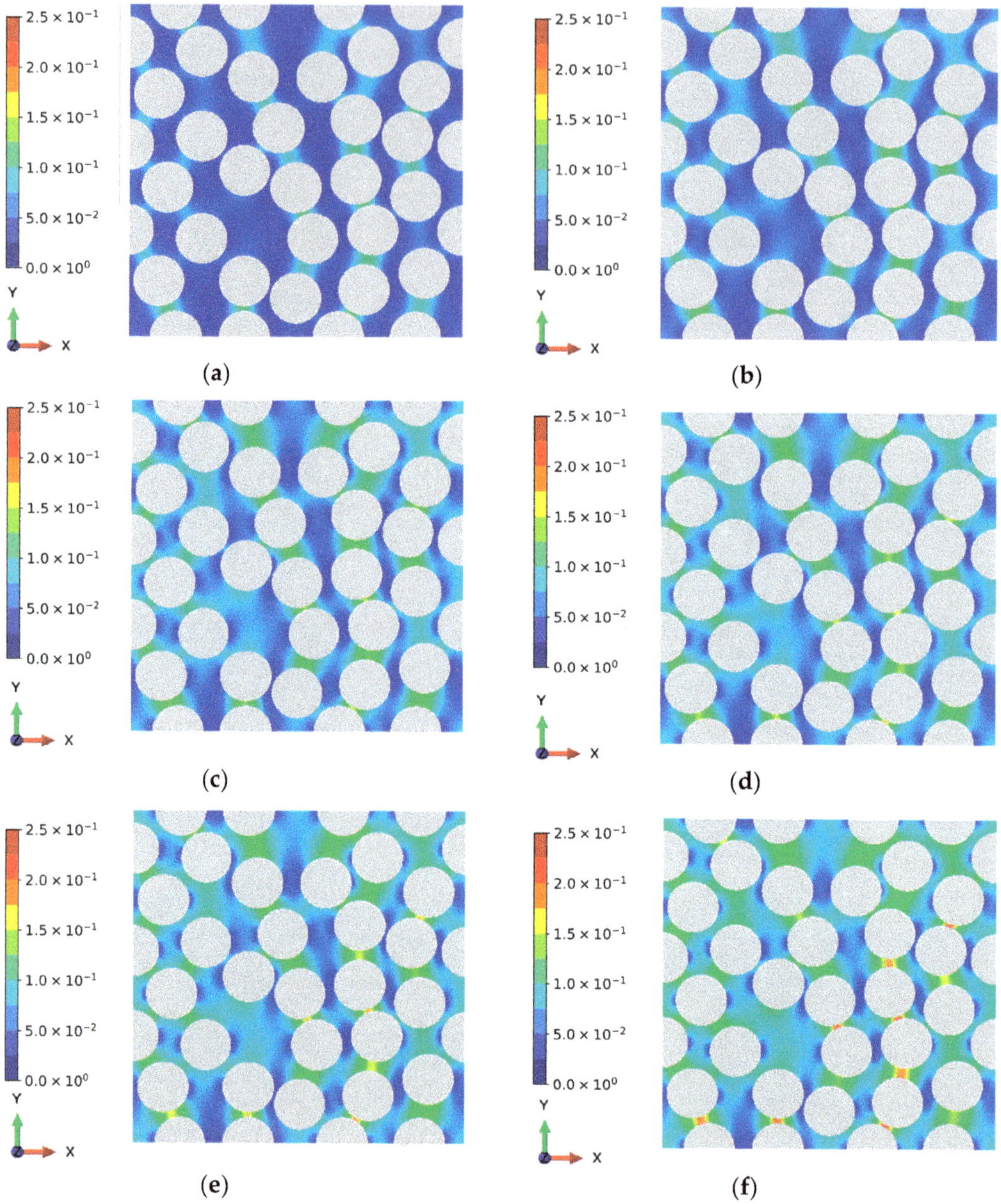

Figure 6. Evolution of damage in CFRPs under high-low cyclic loading sequence. (Palette represents the damage). (**a**) High stress level: 10 cyclic loadings. (**b**) High stress level: 15 cyclic loadings. (**c**) Low stress level: 20 cyclic loadings. (**d**) Low stress level: 30 cyclic loadings. (**e**) Low stress level: 40 cyclic loadings. (**f**) Low stress level: 59 cyclic loadings.

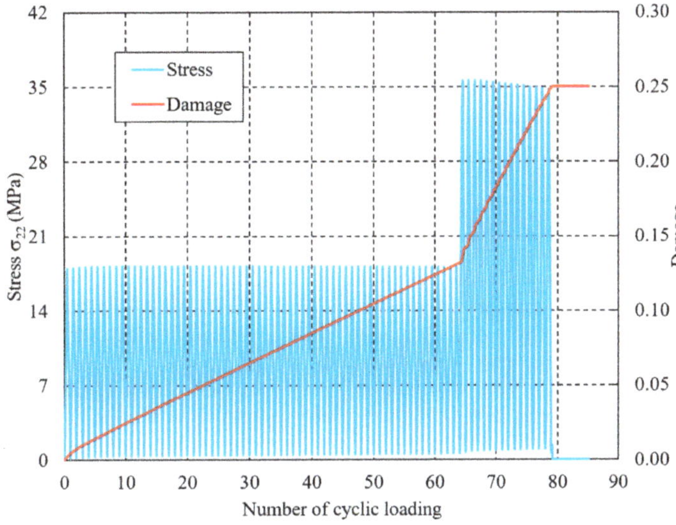

Figure 7. Variations in stress and damage of the element from the initial loading to full failure under the low-high cyclic loading sequence.

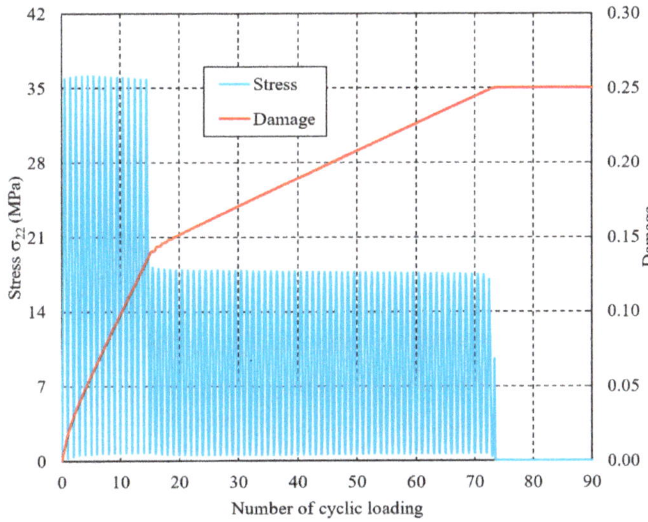

Figure 8. Variations in stress and damage of the element from the initial loading to full failure under the high-low cyclic loading sequence.

By comparing the predicted fatigue lifetimes of the two cases, it can be observed that a significant difference of values, $(79 - 74)/79 = 6.3\%$, occurs. This difference can be utilized to illustrate the dependence of loading sequence on the fatigue lifetimes of components subjected to random loadings [19–21]. To the best knowledge of the authors, the proposed entropy-based failure criterion is the first that has the ability to reproduce the effects of loading sequence on the fatigue lifetime. As the number of cyclic loadings increases, the fatigue lifetime of the H-L loading sequence will be reduced more significantly than the L-H lifetime. It is worth noting that the proposed methodology requires a higher computational cost due to the determination of dashpot elongation in the Maxwell elements. In particular, the computational costs will increase dramatically in the case of cross-ply laminates. In addition to the conventional material properties, the final fracture entropy s^f should also be determined experimentally [52] when dealing with various matrix resins.

3.2. Further Discussions

Before applying the entropy-based failure criterion to estimate the fatigue lifetime under multiple-amplitude cyclic loadings, its validity has been reported by Sato et al. [52], in which only monotonic loading was addressed, and comparisons with experimental results are also conducted. In addition, although the micro-scale model is proposed in this study, it is also appropriate to extend the entropy-based failure criterion to the macro-fatigue failure of CFRPs under cyclic loadings. In fact, Koyanagi et al. [51,53] have adopted the entropy-based failure criterion to investigate the macro-transverse crack accumulation behavior of CFRPs under cyclic loadings. In contrast to the micro-scale modeling, the CFRP is assumed to be a homogeneous orthotropic medium, and only five Maxwell elements are utilized to address the visco-elasticity of the CFRP due to the high computational cost. The well-known Hashin criteria are introduced to characterize the typical failure modes, and the entropy generation is used for material property degradation. The numerical results demonstrate that the entropy-based failure criterion can reveal the effect of the load frequency on the transverse crack accumulation behavior. Thus, the proposed failure criterion is a promising method to predict the residual strengths and fatigue lifetimes of CFRP components.

3.3. Comparisons with LCD and KCG Methodologies

In this section, the fatigue lifetime of the matrix resin is estimated via the conventional linear cumulative damage theory and the kinetic crack growth theory. The failure criterion of LCD [58] is expressed as

$$\int_0^{\tilde{t}} \frac{\tilde{\sigma}^{1/n+1}(\tau)d\tau}{1-\tilde{\sigma}^{1/n}(\tau)} = 1 \tag{12}$$

And the failure criterion of KCG [44] is

$$\frac{1}{1-\tilde{\sigma}\left(\tilde{t}\right)^{1/n}} \int_0^{\tilde{t}} \tilde{\sigma}^{(1/n+1)}(\tau)d\tau = 1 \tag{13}$$

where $\tilde{\sigma}$ and $\tilde{t}$ are the normalized stress and time and the material property parameter is taken as $n = 1.0703$.

Figure 9 compares the variation in damage determined by Equations (12) and (13) versus the number of cyclic loadings in the case of the L-H loading sequence. The damage variable continues increasing after the beginning of cyclic loading, and exhibits differently increasing rates in the various loading levels. It is worth noting that the variation law of the results from LCD is consistent with that from the entropy-based failure criterion. However, significant oscillation phenomenon can be found in the KCG method. The predicted lifetimes of the two methodologies are 80 and 77, respectively, which are in good in agreement with the entropy-based failure criterion's lifetime of 79. In the case of the H-L loading sequence, the variation in damage is shown in Figure 10, and the estimated lifetimes are 81 and 85, respectively.

However, the estimated lifetimes in the case of the H-L loading sequence are always higher than in L-H, regardless of whether the LCD or KCG methodology is used. This observation is contrary to the conclusion discussed in Section 3.1. Additional studies and experiments, which will be presented soon, need to be introduced for a further explanation.

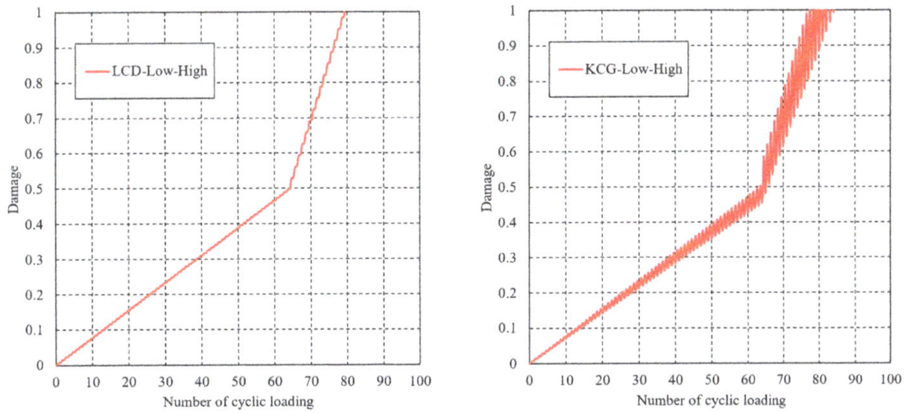

Figure 9. Variations in damage determined by LCD and KCG theories in the L-H loading sequence.

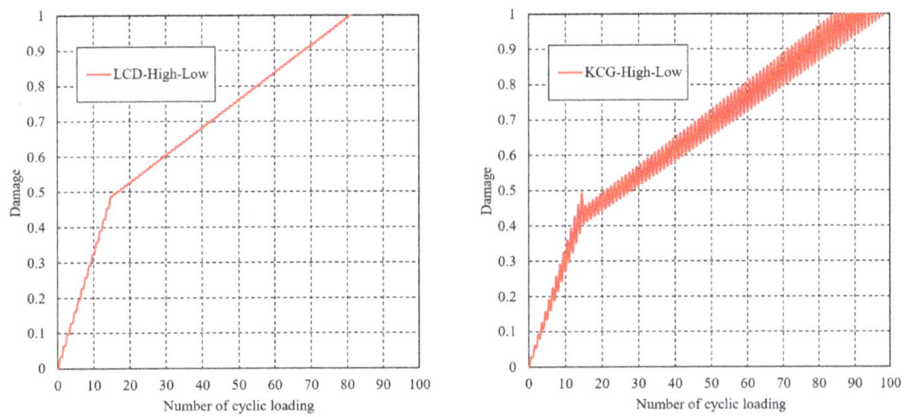

Figure 10. Variations in damage determined by LCD and KCG theories in the H-L loading sequence.

4. Conclusions

The entropy-based failure criterion that can reveal the effect of the loading sequence on the fatigue lifetime is implemented into a UMAT of Abaqus to estimate the fatigue lifetime of unidirectional CFRPs subjected to multiple-amplitude cyclic loadings in this study. Due to the heterogeneity of CFRPs, a micro-finite element model, considering matrix resin and fibers separately, is developed for the numerical simulation. The fatigue lifetimes of CFRPs subjected to L-H and H-L loading sequences are estimated by the proposed method. Numerical results demonstrate that the estimated fatigue lifetime under the L-H loading sequence is higher than that under the H-L loading sequence and the difference in predicted lifetime to final failure is 6.3%. Thus, the proposed method can reveal the effect of the loading sequence on the fatigue lifetime. This is a significant advantage compared to the conventional methods to estimate the residual strengths and lifetimes of CFRP structures. Additionally, comparisons with the LCD and KCG theories are also conducted to illustrate the validity of the proposed method. In future works, the effects of heat generation and thermal conduction on the entropy generation and material property degradation will be included.

Author Contributions: Conceptualization, H.D., K.T., M.S. and J.K.; methodology, K.T. and M.S.; software, K.T. and J.K.; validation, H.D. and K.T.; formal analysis, H.D. and K.T.; investigation, M.S. and J.K.; resources, M.S. and J.K.; data curation, K.T. and J.K.; writing—original draft preparation, H.D. and K.T.; writing—review and editing, H.D. and J.K.; visualization, H.D. and K.T.; supervision, J.K.; project administration, J.K.; funding acquisition, J.K. All authors have read and agreed to the published version of the manuscript.

Funding: This research was funded by JST MIRAI grant number 221036344, and KAKENHI grant number 21KK0063.

Institutional Review Board Statement: Not applicable.

Informed Consent Statement: Informed consent was obtained from all subjects involved in the study.

Data Availability Statement: The data that support the findings of this study are available from the corresponding author upon reasonable request.

Acknowledgments: Parts of this study were financially supported by the JST MIRAI grant number 221036344, and KAKENHI grant number 21KK0063.

Conflicts of Interest: The authors declare no conflict of interest.

Nomenclature

$\boldsymbol{\varepsilon}$, $\Delta\boldsymbol{\varepsilon}$: Total strain tensor, increment of total strain tensor.
$\boldsymbol{\varepsilon}^{ve}$, $\boldsymbol{\varepsilon}^{vp}$: Visco-elasticity strain tensor, visco-plasticity strain tensor.
$\boldsymbol{\sigma}$, $\Delta\boldsymbol{\sigma}$, $\mathbf{s}$: Stress tensor, increment of stress tensor, deviatoric stress tensor.
p, $\mathbf{I}$: Hydrostatic stress, second order identity tensor.
D, D_{cr}: Damage variable, critical damage variable.
s, s^f: Entropy generation, final fracture entropy.
g: Nonlinear coefficient.
t, t', τ: Time.
W, α_d: Dissipated energy, damage parameters.
$\mathbf{E}^r$, $\mathbf{H}$, $\mathbf{M}$: Relaxation tensor, viscosity matrix, constant matrix.
$\mathbf{D}$, E_0, v: Stiffness matrix, initial Young's modulus, Poisson's ratio.
E_n η_n: Visco-elasticity properties of Maxwell elements.
σ_{eqv}, ε_{eqv}^{vp}: Equivalent stress, equivalent visco-plastic strain.
η^{vp}, η_0, α, β, χ: Visco-plasticity coefficients.
$\boldsymbol{A}$, $\boldsymbol{V}$: Generalized thermodynamic force and internal flow vectors.
T, $\boldsymbol{Q}$: Temperature, heat flux vector.

References

1. Kupski, J.; de Freitas, S. Teixeira. Design of adhesively bonded lap joints with laminated CFRP adherends: Review, challenges and new opportunities for aerospace structures. *Compos. Struct.* **2021**, *268*, 113923. [CrossRef]
2. Katafiasz, T.J.; Greenhalgh, E.S.; Allegri, G.; Pinho, S.T.; Robinson, P. The influence of temperature and moisture on the mode I fracture toughness and associated fracture morphology of a highly toughened aerospace CFRP. *Compos. Part A Appl. Sci. Manuf.* **2021**, *142*, 106241. [CrossRef]
3. Lv, T.; Wang, D.; Du, X. Dual-scale parametric modeling and optimal design method of CFRP automotive roof beam. *Compos. Struct.* **2023**, *308*, 116695. [CrossRef]
4. Sim, K.B.; Lee, T.H.; Han, G.Y.; Kim, H.J. Thermal expansion and mechanical properties of urethane-modified epoxy bonded CFRP/steel joints at low and high temperatures for automotive. *Compos. Struct.* **2023**, *322*, 117426. [CrossRef]
5. Pu, Y.; Ma, Z.; Liu, L.; Bai, Y.; Huang, Y. Improvement on strength and toughness for CFRPs by construction of novel "soft-rigid" interface layer. *Compos. Part B Eng.* **2022**, *236*, 109846. [CrossRef]
6. Li, S.-Q.; Liu, R.-G.; Xie, G.-H.; Wang, S.-Y.; Tao, Z.-A. Hygrothermal effect on interfacial properties of CFRP-steel joint modified by nanosilica. *Constr. Build. Mater.* **2023**, *397*, 132318. [CrossRef]
7. Wang, W.; Chen, X.; Fan, H. Modular technique to construct lightweight CFRP lattice structures. *Thin-Walled Struct.* **2023**, *182*, 110259. [CrossRef]
8. Iqbal, S.; Jamil, T.; Mehdi, S.M. Numerical simulation and validation of MWCNT-CFRP hybrid composite structure in lightweight satellite design. *Compos. Struct.* **2023**, *303*, 116323. [CrossRef]

9. Hazell, P.J.; Stennett, C.; Cooper, G. The effect of specimen thickness on the shock propagation along the in-fibre direction of an aerospace-grade CFRP laminate. *Compos. Part A Appl. Sci. Manuf.* **2009**, *40*, 204–209. [CrossRef]
10. Ning, H.; Li, Y.; Li, J.; Hu, N.; Liu, Y.; Wu, L.; Liu, F. Toughening effect of CB-epoxy interleaf on the interlaminar mechanical properties of CFRP laminates. *Compos. Part A Appl. Sci. Manuf.* **2015**, *68*, 226–234. [CrossRef]
11. Vassilopoulos, A.P. The history of fiber-reinforced polymer composite laminate Fatigue. *Int. J. Fatigue* **2020**, *134*, 105512. [CrossRef]
12. Schapery, R.A. On the characterization of nonlinear viscoelastic materials. *Polym. Eng. Sci.* **1969**, *9*, 295–310. [CrossRef]
13. Yokozeki, T.; Aoki, T.; Ishikawa, T. Fatigue growth of matrix cracks in the transverse direction of CFRP laminates. *Compos. Sci. Technol.* **2002**, *62*, 1223–1229. [CrossRef]
14. Deng, H.; Yan, B.; Zhang, X.; Zhu, Y. A new enrichment scheme for the interfacial crack modeling using the XFEM. *Theor. Appl. Fract. Mech.* **2022**, *122*, 103595. [CrossRef]
15. Deng, H.; Yan, B.; Okabe, T. Fatigue crack propagation simulation method using XFEM with variable-node element. *Eng. Fract. Mech.* **2022**, *269*, 108533. [CrossRef]
16. Deng, H.; Yan, B.; Zhang, X.; Zhu, Y.; Koyanagi, J. New crack front enrichment for XFEM modeling. *Int. J. Solids Struct.* **2023**, *274*, 112280. [CrossRef]
17. Panella, F.W.; Pirinu, A. Fatigue and damage analysis on aeronautical CFRP elements under tension and bending loads: Two cases of study. *Int. J. Fatigue* **2021**, *152*, 106403. [CrossRef]
18. Ferreira, J.A.M.; Reis, P.N.; Costa, J.D.M.; Richardson, M.O.W. Fatigue behaviour of composite adhesive lap joints. *Compos. Sci. Technol.* **2002**, *62*, 1373–1379. [CrossRef]
19. Lin, J.X.; Huang, P.Y.; Guo, Y.C.; Guo, X.Y.; Zeng, J.J.; Zhao, C.; Chen, Z.B. Fatigue behavior of RC beams strengthened with CFRP laminate under hot-wet environments and vehicle random loads coupling. *Int. J. Fatigue* **2020**, *131*, 105329. [CrossRef]
20. Hosoi, A.; Kawada, H.; Yoshino, H. Fatigue characteristics of quasi-isotropic CFRP laminates subjected to variable amplitude cyclic two-stage loading. *Int. J. Fatigue* **2006**, *28*, 1284–1289. [CrossRef]
21. Chen, Z.; Huang, P.; Yao, G.; Guo, X.; Yang, Y.; Li, W.; Wu, B. Experimental study on fatigue performance of RC beams strengthened with CFRP under variable amplitude overload and hot-wet environment. *Compos. Struct.* **2020**, *244*, 112308. [CrossRef]
22. Kapidžić, Z.; Granados, D.L.Á.; Arias, J.A.M.; Aguilera, M.J.Q.; Rodríguez, J.P.C.; Callejas, J.C.G. Bolt fatigue in CFRP joints. *Int. J. Fatigue* **2022**, *164*, 107138. [CrossRef]
23. Xie, G.-H.; Tang, Y.-S.; Wang, C.; Li, S.-Q.; Liu, R.-G. Experimental study on fatigue performance of adhesively bonded anchorage system for CFRP tendons. *Compos. Part B Eng.* **2018**, *150*, 47–59. [CrossRef]
24. D'Amore, A.; Giorgio, M.; Grassia, L. Modeling the residual strength of carbon fiber reinforced composites subjected to cyclic loading. *Int. J. Fatigue* **2015**, *78*, 31–37. [CrossRef]
25. Hosoi, A.; Arao, Y.; Karasawa, H.; Kawada, H. High-cycle fatigue characteristics of quasi-isotropic CFRP laminates. *Adv. Compos. Mater.* **2007**, *16*, 151–166. [CrossRef]
26. Reifsnider, K.L.; Talug, A. Analysis of fatigue damage in composite laminates. *Int. J. Fatigue* **1980**, *2*, 3–11. [CrossRef]
27. Ogihara, S.; Takeda, N.; Kobayashi, S.; Kobayashi, A. Effects of stacking sequence on microscopic fatigue damage development in quasi-isotropic CFRP laminates with interlaminar-toughened layers. *Compos. Sci. Technol.* **1999**, *59*, 1387–1398. [CrossRef]
28. Li, C.; Guo, R.; Xian, G.; Li, H. Effects of elevated temperature, hydraulic pressure and fatigue loading on the property evolution of a carbon/glass fiber hybrid rod. *Polym. Test.* **2020**, *90*, 106761. [CrossRef]
29. Miyano, Y.; Nakada, M.; Mcmurray, M.K.; Muki, R. Prediction of flexural fatigue strength of CRFP composites under arbitrary frequency, stress ratio and temperature. *J. Compos. Mater.* **1997**, *31*, 619–638. [CrossRef]
30. Xiao, X.R. Modeling of load frequency effect on fatigue life of thermoplastic composites. *J. Compos. Mater.* **1999**, *33*, 1141–1158. [CrossRef]
31. Tsai, S.W. *Composite Materials: Testing and Design (Fifth Conference)*; ASTM International: Conshohocken, PA, USA, 1979.
32. Vitale, P.; Francucci, G.; Rapp, H.; Stocchi, A. Manufacturing and compressive response of ultra-lightweight CFRP cores. *Compos. Struct.* **2018**, *194*, 188–198. [CrossRef]
33. Found, M.S.; Lamb, J.R.; Moore, P.; Jones, M.W. Comparison of damage resistance of CFRP lightweight panels. *Compos. Part A Appl. Sci. Manuf.* **2005**, *36*, 197–203. [CrossRef]
34. Hua, X.; Higuchi, R.; Yokozeki, T. Enhancement of tensile strength of tapered laminates by utilizing thin-ply composites. *Compos. Part B Eng.* **2023**, *248*, 110372. [CrossRef]
35. Chen, Y.; Li, M.; Yang, X.; Luo, W. Damage and failure characteristics of CFRP/aluminum single lap joints designed for lightweight applications. *Thin-Walled Struct.* **2020**, *153*, 106802. [CrossRef]
36. Zhang, J.; An, Q. Topology optimization of fibre reinforced polymer lattice structures for additive manufacturing. *Compos. Sci. Technol.* **2023**, *242*, 110144. [CrossRef]
37. Sun, Q.; Zhou, G.; Meng, Z.; Jain, M.; Su, X. An integrated computational materials engineering framework to analyze the failure behaviors of carbon fiber reinforced polymer composites for lightweight vehicle applications. *Compos. Sci. Technol.* **2021**, *202*, 108560. [CrossRef] [PubMed]
38. Asp, L.E.; Berglund, L.A.; Talreja, R. Prediction of matrix-initiated transverse failure in polymer composites. *Compos. Sci. Technol.* **1996**, *56*, 1089–1097. [CrossRef]
39. Gosse, J.; Christensen, S. *Strain Invariant Failure Criteria for Polymer in Composite Materials*; AIAA: Reston, VA, USA, 2001; p. 1184.

40. Tay, T.E.; Tan, S.H.N.; Tan, V.B.C.; Gosse, J.H. Damage progression by the element-failure method (EFM) and strain invariant failure theory (SIFT). *Compos. Sci. Technol.* **2005**, *65*, 935–944. [CrossRef]
41. Canal, L.P.; Segurado, J.; LLorca, J. Failure surface of epoxy-modified fiber-reinforced composites under transverse tension and out-of-plane shear. *Int. J. Solids Struct.* **2009**, *46*, 2265–2274. [CrossRef]
42. McCartney, L.N. Crack propagation, resulting from a monotonic increasing applied stress, in a linear viscoelastic material. *Int. J. Fract.* **1977**, *13*, 641–654. [CrossRef]
43. Cartney, L.N. Crack growth laws for a variety of viscoelastic solids using energy and COD fracture criteria. *Int. J. Fract.* **1979**, *15*, 31–40. [CrossRef]
44. Christensen, R.M. An evaluation of linear cumulative damage (Miner's law) using kinetic crack growth theory. *Mech. Time-Dependent Mater.* **2002**, *6*, 363–377. [CrossRef]
45. Christensen, R.; Miyano, Y. Stress intensity controlled kinetic crack growth and stress history dependent life prediction with statistical variability. *Int. J. Fract.* **2006**, *137*, 77–87. [CrossRef]
46. Koyanagi, J.; Sato, Y.; Sasayama, T.; Okabe, T.; Yoneyama, S. Numerical simulation of strain-rate dependent transition of transverse tensile failure mode in fiber-reinforced composites. *Compos. Part A Appl. Sci. Manuf.* **2014**, *56*, 136–142. [CrossRef]
47. Haghshenas, A.; Jang, J.Y.; Khonsari, M.M. On the intrinsic dissipation and fracture fatigue entropy of metals. *Mech. Mater.* **2021**, *155*, 103734. [CrossRef]
48. Naderi, M.; Amiri, M.; Khonsari, M.M. On the thermodynamic entropy of fatigue fracture. *Proc. R. Soc. A Math. Phys. Eng. Sci.* **2010**, *466*, 423–438. [CrossRef]
49. Huang, J.; Li, C.; Liu, W. Investigation of internal friction and fracture fatigue entropy of CFRP laminates with various stacking sequences subjected to fatigue loading. *Thin-Walled Struct.* **2020**, *155*, 106978. [CrossRef]
50. Huang, J.; Yang, H.; Liu, W.; Zhang, K.; Huang, A. Confidence level and reliability analysis of the fatigue life of CFRP laminates predicted based on fracture fatigue entropy. *Int. J. Fatigue* **2022**, *156*, 106659. [CrossRef]
51. Koyanagi, J.; Mochizuki, A.; Higuchi, R.; Tan, V.B.C.; Tay, T.E. Finite element model for simulating entropy-based strength-degradation of carbon-fiber-reinforced plastics subjected to cyclic loadings. *Int. J. Fatigue* **2022**, *165*, 107204. [CrossRef]
52. Sato, M.; Hasegawa, K.; Koyanagi, J.; Higuchi, R.; Ishida, Y. Residual strength prediction for unidirectional CFRP using a nonlinear viscoelastic constitutive equation considering entropy damage. *Compos. Part A Appl. Sci. Manuf.* **2021**, *141*, 106178. [CrossRef]
53. Deng, H.; Mochizuki, A.; Fikry, M.; Abe, S.; Ogihara, S.; Koyanagi, J. Numerical and Experimental Studies for Fatigue Damage Accumulation of CFRP Cross-Ply Laminates Based on Entropy Failure Criterion. *Materials* **2023**, *16*, 388. [CrossRef] [PubMed]
54. Basaran, C.; Nie, S. An irreversible thermodynamics theory for damage mechanics of solids. *Int. J. Damage Mech.* **2004**, *13*, 205–223. [CrossRef]
55. González, C.; LLorca, J. Mechanical behavior of unidirectional fiber-reinforced polymers under transverse compression: Microscopic mechanisms and modeling. *Compos. Sci. Technol.* **2007**, *67*, 2795–2806. [CrossRef]
56. Li, S.; Warrior, N.; Zou, Z.; Almaskari, F. A unit cell for FE analysis of materials with the microstructure of a staggered pattern. *Compos. Part A Appl. Sci. Manuf.* **2011**, *42*, 801–811. [CrossRef]
57. Melro, A.; Camanho, P.; Pires, F.A.; Pinho, S. Micromechanical analysis of polymer composites reinforced by unidirectional fibres: Part II-Micromechanical analyses. *Int. J. Solids Struct.* **2013**, *50*, 1906–1915. [CrossRef]
58. Miner, M.A. Cumulative Damage in Fatigue. *J. Appl. Mech.* **1945**, *12*, A159–A164. [CrossRef]

materials

MDPI

Article

Damage Assessment of Glass-Fibre-Reinforced Plastic Structures under Quasi-Static Indentation with Acoustic Emission

Norman Osa-uwagboe [1,2], Amadi Gabriel Udu [2,3], Vadim V. Silberschmidt [1,*], Konstantinos P. Baxevanakis [1] and Emrah Demirci [1]

1 Wolfson School of Mechanical, Electrical, and Manufacturing Engineering, Loughborough University, Loughborough LE11 3TU, UK; n.osa-uwagboe@lboro.ac.uk (N.O.-u.); k.baxevanakis@lboro.ac.uk (K.P.B.); e.demirci@lboro.ac.uk (E.D.)
2 Air Force Research and Development Centre, Nigerian Air Force Base, Kaduna 800282, Nigeria; agu1@leicester.ac.uk
3 School of Engineering, University of Leicester, Leicester LE1 7RH, UK
* Correspondence: v.silberschmidt@lboro.ac.uk; Tel.: +44-(0)-1509-227504

Abstract: The use of fibre-reinforced plastics (FRPs) in various industrial applications continues to increase thanks to their good strength-to-weight ratio and impact resistance, as well as the high strength that provides engineers with advanced options for the design of modern structures subjected to a variety of out-of-plane impacts. An assessment of the damage morphology under such conditions using non-destructive techniques could provide useful data for material design and optimisation. This study investigated the damage mechanism and energy-absorption characteristics of E-glass laminates and sandwich structures with GFRP face sheets with PVC cores under quasi-static indentation with conical, square, and hemispherical indenters. An acoustic emission (AE) technique, coupled with a k-means++ pattern-recognition algorithm, was employed to identify the dominant microscopic and macroscopic damage mechanisms. Additionally, a post-mortem damage assessment was performed with X-ray micro computed tomography and scanning electron microscopy to validate the identified clusters. It was found that the specific energy absorption after impact with the square and hemispherical indenters of the GFRP sandwich and the plain laminate differed significantly, by 19.29% and 43.33%, respectively, while a minimal difference of 3.5% was recorded for the conical indenter. Additionally, the results obtained with the clustering technique applied to the acoustic emission signals detected the main damaged modes, such as matrix cracking, fibre/matrix debonding, delamination, the debonding of face sheets/core, and core failure. The results therefore could provide a methodology for the optimisation and prediction of damage for the health monitoring of composites.

Keywords: quasi-static indentation; fibre-reinforced plastics; acoustic emission; damage; X-ray micro computed tomography; scanning electron microscopy

Citation: Osa-uwagboe, N.; Udu, A.G.; Silberschmidt, V.V.; Baxevanakis, K.P.; Demirci, E. Damage Assessment of Glass-Fibre-Reinforced Plastic Structures under Quasi-Static Indentation with Acoustic Emission. *Materials* **2023**, *16*, 5036. https://doi.org/10.3390/ma16145036

Academic Editor: Luís Miguel Pereira Durão

Received: 30 June 2023
Revised: 12 July 2023
Accepted: 14 July 2023
Published: 17 July 2023

1. Introduction

Composite materials have been used increasingly in various industrial applications in the past 50 years thanks to their inherent advantage of a greater stiffness-to-weight ratio than that of traditional metals. A popular variant 36 of composite materials is fibre-reinforced plastic sandwich structures (FRPSS), made from thermosetting resins due to their ease of manufacture, buoyancy, and acceptable impact resistance performance [1,2]. These materials are commonly made of glass fibre-reinforced plastics (GFRP) or carbon fibre-reinforced plastics (CFRP), and foam cores, whose performances are also affected by moisture ingress. However, the overall performance of FRPSSs makes them suitable for the construction of both above-water and underwater marine vessels [3–5]. These applications can expose the structures to low-velocity impact (LVI), which can produce barely visible impact damage (BVID) and lead to catastrophic failure. Several studies have adopted different measures for BVID in composite materials in order to comprehend the damage modes pertinent to these types of loading conditions [6]. Owing to the similarities in damage scenarios, it was demonstrated that a quasi-static indentation (QSI) method could give indications of the damage modes in LVI for composites [7,8]. Zniker et al. used this method to compare the energy-absorption capability of GFRP laminates and PVC-foam sandwich structures under repeated impacts and the reduced levels of energy using a modified Charpy test and QSI experiments [9]. The obtained results revealed that while the indentation energy for the laminates with varying thicknesses was identical, the presence of the foam core significantly improved the damping properties of the sandwich structures. Furthermore, the damaged area in the sandwich structures was larger than that of the laminates and predominantly characterised by delamination. Other studies showed that failure modes of sandwich structures due to QSI are similar to those caused by LVI and include core buckling, core crushing, delamination in the face sheets, and debonding between the core and face sheets, as well as matrix cracking and fibre breakage in the face sheets [10–13].

Considering that damage evolution is a critical component of structural health monitoring in composites, several non-destructive techniques (NDT), such as acoustic emission (AE), were adopted because they can monitor internal damage and provide useful information on the damage evolution process in real time [14–16]. Dikshit et al. successfully utilised AE techniques to characterise the energy-absorption properties of FRPSSs with varying core designs. The results showed that various damage modes (fibre breakage, fibre/matrix debonding) were obtained, and a correlation between the AE features and the corresponding absorption properties of the various cores was established [17]. Hajikhani et al. applied the AE technique to assess the strain energy release rate in mode I delamination in FRPSS with woven GFRP face sheets and a foam core. It was shown that AE signals coupled with results from mechanical tests accurately described the fracture toughness, as well as the in-plane fibre orientation effect on the fracture behaviour [18]. Further, Ben Ammar et al. used AE coupled with clustering techniques to identify and characterise the local damage in FRPSS with PVC closed-cell foam cores with different densities under a quasi-static loading regime; the obtained results demonstrated a good agreement with the experimental data [19]. Other studies also adopted the approach of utilising AE signals and clustering or pattern recognition techniques for damage analysis in composite materials [20–23]. Hence, the AE technique and the cluster analysis are a reliable methodology for the monitoring and evaluation of the damage-evolution process and the recognition of the distinctive damage modes. However, it is worth noting that a better understanding of the internal structural characteristics can be obtained post mortem, using other NDT techniques. A common approach is the use of X-ray micro computed tomography (μ-CT), as demonstrated in [15,24]. Therefore, a combination of the above processes can provide a clearer picture of the damage sequence in FRPSSs and an understanding of the phenomenological progression of damage. For instance, [25] investigated the indentation response of sandwich structures to various indenter geometries and was able to capture the constituent failure modes accurately.

To gain a broader perspective for optimising the component design and ongoing structural health monitoring, it is important to investigate the damage mechanisms and energy-absorption capabilities of composite materials, especially laminates and sandwich structures. Thus, in this paper, a machine-learning (ML) model was applied to analyse the AE features, supporting a comprehensive understanding of the damage characteristics of GFRP laminates and sandwich structures. The proposed methodology could provide valuable insights into the optimisation of composites and prediction of damage with their health monitoring in real-life applications. To this end, this study compared the damage sequence of GFRP laminates (GL) and GFRP sandwich panels (GS) using AE to identify their damage modes.

2. Experimental Procedure

A schematic of the experimental setup used in this study is presented in Figure 1. As stated, two different sample types, GL and GS, were tested and subjected to QSI with three different indenter shapes, monitored with AE to capture visible and invisible damage in the specimens. The average data were obtained and damage assessment was conducted with a clustering algorithm, as well as μ-CT/SEM investigations.

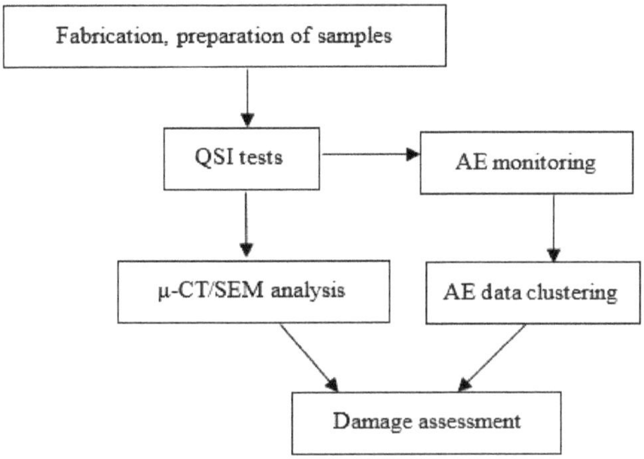

Figure 1. Schematic of experimental setup.

2.1. *Materials and Manufacturing Methods*

Aerospace-grade E-glass plain weave fabric with 160 g/m^2 from Samson Composites Ltd. (Shenzhen, China) was used as reinforcement and face sheets for the FRPSS, while epoxy resin with C-1 catalyst hardener from EPOCHEM Ltd. (Lagos, Nigeria) with a volumetric ratio of 2:1 was employed as the matrix system to fabricate samples at room temperature (27 °C in Nigeria) with a curing time of 18 h. For the core, EASYCell 75 closed-cell PVC foam core forms from EASY composites (Stoke-on-Trent, UK) were used [26]. Details of the constituents' parameters are given in Table 1. A hand lay-up and a vacuum-bagging technique, as described in Figure 2, were deployed to produce plates of 300 mm × 300 mm with a configuration of 8 layers of E-glass fabric for the GL samples and 4 layers/PVC foam/4 layers for the GS specimen. In order to reduce the presence of air bubbles in the mixture of the hardener and resin, it was necessary to degas the mixture for about 2 min before applying it to the sample and subsequent vacuum bagging. This mode of fabrication was selected due to its low cost, simplicity, and wide range of applicability.

Table 1. Mechanical properties of constituent materials [27–29].

Material	Young's Modulus (GPa)	Shear Modulus (GPa)	Tensile Strength (MPa)	Poisson Ratio	Density (g/cm^3)
	E	G_{12}		ν_{12}	ρ
E-glass fabric	72.39	8.27	3100–3800	0.26	2.25
PVC foam	0.075	0.028	1.89	-	0.075
Epoxy matrix	3.2–3.5	-	70–80	0.29	1.16

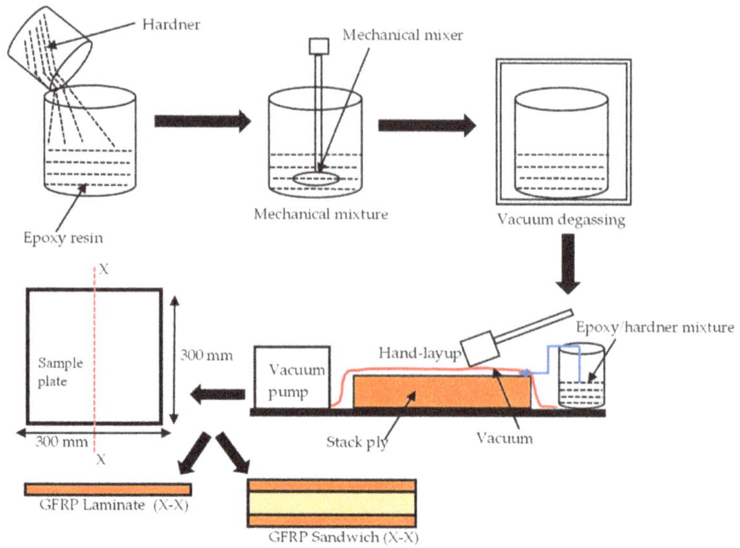

Figure 2. Schematic of sample fabrication process.

2.2. *Quasi-Static Indentation Tests*

QSI was performed on the samples with different indenter shapes to study the damage tolerance of the materials, to provide useful data on the sequence of damage, with a minimum of 5 samples per configuration. The experiment was carried out in line with the ASTM D6264/D6264M-17 standard test method with a displacement control of 1 mm/min, while the vertical displacement was measured using a linear variable differential transformer (LVDT) [30]. The test was performed using an Instron 3369 Universal machine 50 kN cell (Instron Corporation, Norwood, MA, USA) with a fixture made from steel plates. Three different indenter types made from stainless steel were used for this experiment: hemispherical, conical, and flat indenters with a diameter/length of 13 mm. During the test, the indenter was aligned to the centre of the specimen with an offset at no more than 0.01 mm, and then indented until complete perforation. The experimental jig setup is depicted in Figure 3, while the indenter geometry is shown in Figure 4.

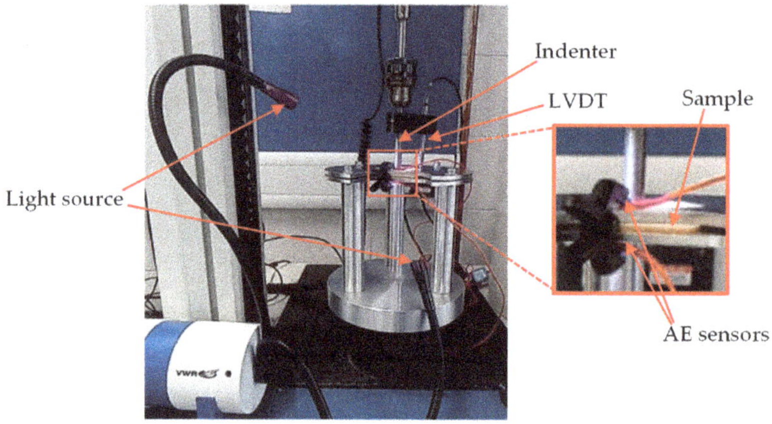

Figure 3. QSI experimental setup.

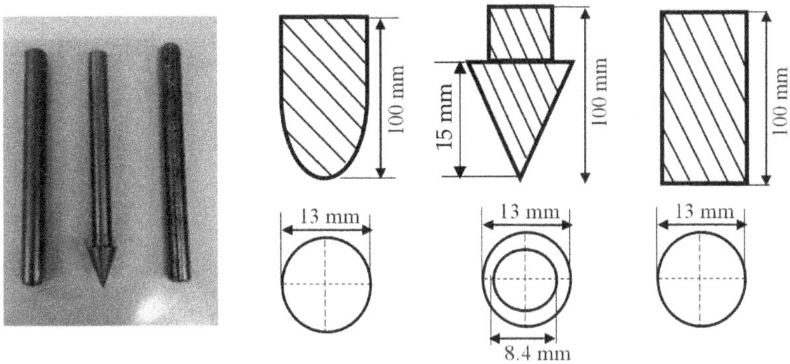

Figure 4. Indenter types and dimensions.

Under out-of-plane loading conditions, energy absorption is an important parameter to understand the damage evolution in composites. Therefore, a quantitative comparison of the energy-absorption properties of the samples under various indenter configurations was performed based on the following expression:

$$E_a = \int_{x_0}^{x_1} F(x)dx \quad (1)$$

where E_a is the total energy absorbed, obtained by integration of the area under the force-displacement curve. Thereafter, the specific energy absorption (SEA), which relates to the energy absorbed per area density (A), could be calculated as follows:

$$SEA = \frac{E_a}{A} \quad (2)$$

All samples were codified depending on the indenter geometries and the loading regimes for the ease of identification. For example, the specimens denoted by GLH and GSC represent glass laminate indented by hemispherical indenter, and glass sandwich indented by a conical punch, respectively. The complete specimen nomenclature is given in Table 2.

Table 2. Nomenclature of samples.

Designation	Thickness (mm)	Description
GLH	2	Laminate/hemispherical punch
GLC	2	Laminates/conical punch
GLS	2	Laminate/square punch
GSH	5	Sandwich/hemispherical punch
GSC	5	Sandwich/conical punch
GSS	5	Sandwich/square punch

2.3. *Acoustic Emission*

2.3.1. Experimental Setup

In total, 18 signal parameters, including the time and frequency domain features, were retrieved from the AE. These include Time (s), Class ID, Channel, Parametric, Risetime, Counts to Peak, Counts, Energy (J), Duration (s), Amplitude (dBae), ASL, Threshold, Average Frequency (Hz), RMS, Signal Strength, Absolute Energy (J), Frequency Centroid, and Peak Frequency (Hz). A typical description of an AE waveform and relevant parameters are shown in Figure 5. These features were utilised by the ML algorithm to assess the damage. A three-step approach was adopted for this task, namely, data standardisation, feature selection, and clustering analysis, as shown in Figure 6. A Micro-SHM system with a frequency band of 1 kHz–1 MHz (Physical Acoustics Corporation, Township, NJ, USA) with 4 AE channels and 2 parametric channels was used. However, only two channels connected to the Nano-30 AE sensors (125 kHz–750 MHz), mounted on the top and bottom face sheets (Figure 3) were used in this study, and the peak amplitude vs. time was obtained.

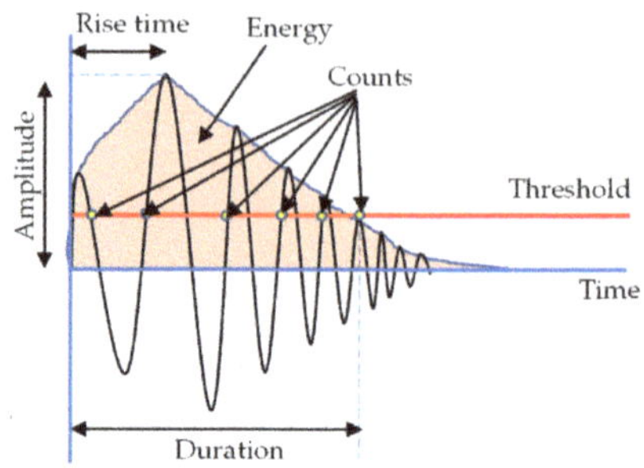

Figure 5. AE signal parameters.

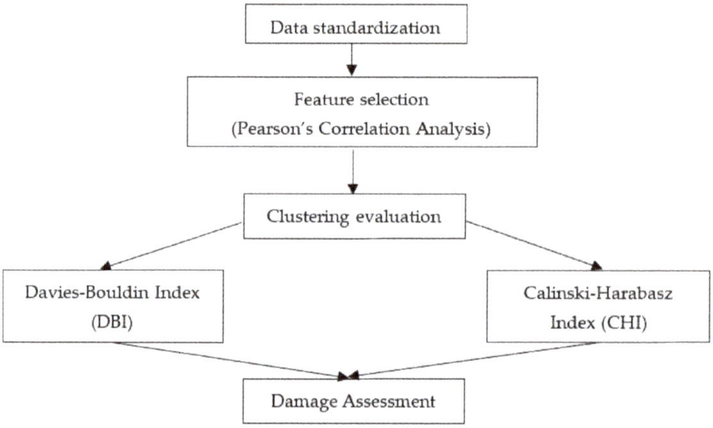

Figure 6. ML framework for damage assessment.

2.3.2. Data Standardisation, Feature Selection, and Cluster Analysis

In order to exclude the dominance problems and prevent calculation complexities for the features retrieved from the AE signal parameters, data standardisation was undertaken prior to other ML tasks. All features (i.e., $x = (x_1, \ldots, x\_n)^T \epsilon R^{(N \, x \, M)}$) were centred and scaled independently, with the standard score sample calculated using $z = (x - \mu)/\sigma$, where μ and σ are the mean and the standard deviation of the training data while x represents the original feature value of the matrix X. This ensured that the means were within a (0–1) range. Features can be categorised as relevant, irrelevant, and redundant. Accordingly, the inclusion/exclusion of features influences the performance of the ML algorithm. Generally, it is a challenge to ascertain which features would optimise ML algorithms' performances for damage assessment. Hence, the feature selection techniques were employed with a view to choosing a subset from an original variable that best represents the underlying pattern, concept, or constructs investigated in the analysis. By reducing/excluding redundant and irrelevant features, ML algorithms are less likely to be misled into making a decision based on noise. Also, removing irrelevant features circumvents overfitting problems, reduces the computational cost, and enhances model accuracy [31–33]. In this study, feature selection was utilised to determine the quality of clustering through expert judgment and Pearson's correlation coefficient (PCC). Correlation between data points x and y is the measure of the linear relationship between 2 features, A and B. PCC takes values ranges of ±1, where a value of zero represents the lack of linear correlation while −1 and +1 stand for perfect negative and positive correlation, respectively. PCC is adept in feature selection since it is based on the method of covariance and was employed in the literature for damage analysis [34,35]. PCC is defined as:

$$r = \frac{\sum_{i=1}^{n}(x_i - \overline{x})(y_i - \overline{y})}{\sqrt{\sum_{i=1}^{n}(x_i - \overline{x})^2 \sum_{i=1}^{n}(y_i - \overline{y})^2}} \quad (3)$$

where n is the total number of samples, while $\overline{x}$ and $\overline{y}$ are the average values of input data x_i and y_i, respectively. Furthermore, this study adopted peak frequency and the amplitude for subsequent damage assessment. Previous studies [35,36] showed that some AE features provided more reliable data for assessing damage, with the peak frequency being the most appropriate since it is not likely to be affected by attenuation, while the amplitude was computed to have the lowest PCC score among other time-domain features. Therefore, out of the 18 features, only 7 were analysed using AE.

Two cluster validity indices, the Calinski–Harabasz index (CHI) and the Davies–Bouldin index (DBI), were adopted to estimate the number of clusters for damage assessment. These indices were combined in previous studies in estimating the optimal number of clusters for damage assessment in AE [14,35,37]. CHI gives a ratio of the degree of separation between clusters to the degree of inter-cluster dispersion, with higher CHI scores suggesting better results of well-separated and tightly packed clusters. Mathematically, CHI is defined as:

$$CHI = \frac{B}{W} \cdot \frac{n_X - k}{k - 1} \quad (4)$$

where B and W are the between and inter-cluster variances, respectively, while $_X$ is the data of size n_X clustered into k. Based on DBI, well-separated clusters with low intra-cluster variance have higher scores than tightly packed clusters with high intra-cluster variance. DBI is defined as:

$$DBI = \frac{1}{k} \sum_{m=1}^{k} \max_{n \neq m} \left(\frac{P_m + P_n}{D_{mn}} \right) \quad (5)$$

where k is the number of clusters, while D_{mn} is the distance between the centroids of clusters m and n; $P_{(m,n)}$ is the average distance between all points in cluster m and the centroid of cluster n. Low DBI implies a dense and well-separated cluster. Furthermore, this study employed the k-means++, an unsupervised learning algorithm that minimised the distance between vectors in the cluster. The algorithm randomly selects a single point from the AE dataset as the first cluster centre and iteratively selects the remaining cluster centres by choosing new centres that are far away from the previously chosen ones. Once all defined numbers of centres have been selected, the algorithm assigns each data point to its nearest centre and recalculates the centres based on the new assignments. This process is repeated until there is no significant change in assigning data points to clusters. The optimum number of clusters is indicated by the cluster validity combined with the least separation. It is worth noting that AE signals comprise volumetric information for deformation and damage in the specimens. A one-to-one correlation of acoustic signals with specific damage events and location was not attempted in this study.

2.4. Damage Characterisation

2.4.1. X-ray μ-CT

The samples were scanned using a high-resolution X-ray μ-CT system NIKON XTH X-Tex 160Xi, (NIKON Metrology Europe, Leuven, Belgium) with an effective pixel size of 27.79 μm. The beam energy, beam current, and power settings were 65 kV, 65 μA, and 4.2 W, respectively, with an exposure time of 500 ms. A total of 3016 tiff images were created per sample scan, with the region of interest limited to the areas surrounding the damaged portion of the samples, as can be seen in the experimental setup in Figure 7. The acquired microscopy data were first processed with VG Studio Max 3.1 software (Volume Graphics, Charlotte, NC, USA) and then post-processed using commercial software (Dragonfly ORS, Adelaide, South Australia, Australia). The optimum centre of rotation was determined in the initial post-processing phase, with all generated slices combined to develop the volumetric image in the reconstruction phase. Gaussian filters were used to reduce the characterisation noise generated by the volumetric imaging, which led to a minimal data loss from the sample materials' homogeneity. The generated tiff files were incorporated into the Dragonfly software with voxel analysis to determine damage parameters.

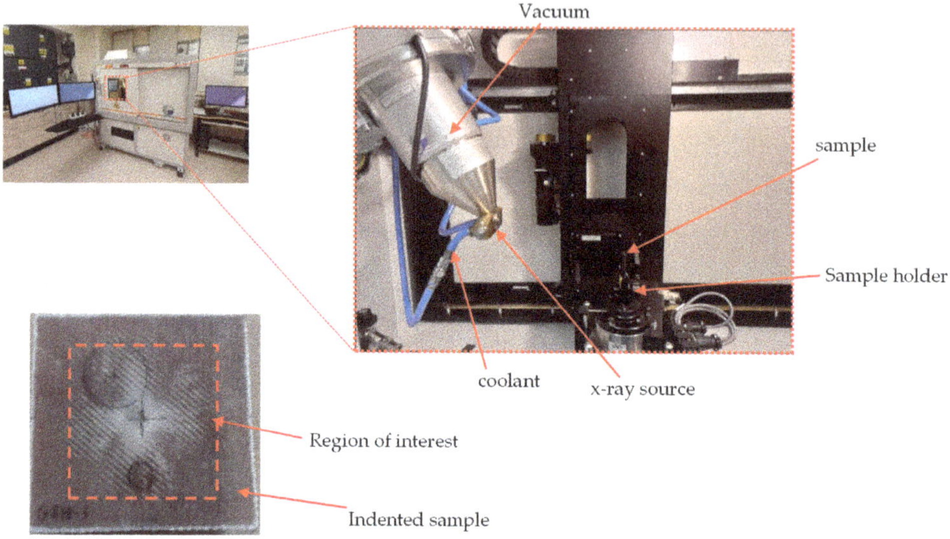

Figure 7. X-ray μ-CT experimental setup.

2.4.2. SEM Analysis

Before conducting SEM analysis, the samples were sputter-coated with Au/Pd with a coating thickness of about 6 nm, then images were captured with a JSM-7800F SEM machine (UK) at an acceleration voltage of 10 kV and probe current of 200 pA. Image post-processing was carried out using Aztec software (Oxford Instruments, Oxford, UK), and the damage mechanism was accurately identified.

3. Results and Discussion

3.1. *Load–Displacement Results*

The average displacement curve for the GS and GL samples for the different indenter geometries are shown in Figure 8. In general, the samples exhibited quasi-brittle failure in three identifiable stages, with two noticeable peaks for the GS samples and a single peak for the GL samples in the load–displacement graphs. These peaks corresponded to the failure of the top and bottom face sheets in the GS samples and laminate failure in the GL specimen. A similar trend of peak forces in the load displacement curves representing the damage characteristics of FRP were reported in [23,24]. For the GS samples described in Figure 8a, it was observed that the load increased in a similar pattern up to the fracture of the bottom face sheets at 2.67 kN, 2.17 kN, and 0.58 kN for GSS, GSH, and GSC, respectively. This variation could be attributed to the difference in the surface area in contact with the sample, as a reduction in the load was required for the penetration of the top face sheet as the indenter angle in contact with the specimen reduced. Consequently, a smaller force was necessary for penetration as the indenter progressed through the sample thickness. The GSC specimen exhibited the top face sheet penetration at a displacement of 1.5 mm, representing 28.3% and 26.3% of the displacement required for the corresponding face sheet penetration in GSH and GSS. Additionally, the penetration load was significantly lower at 79.9% and 85.9% of the force required for GSH and GSS, respectively. After the drop in the load at the first peak, there was a rise in load due to the contribution of the bottom face sheet resistance and the friction of the core as the indenter progressed through the thickness of the sample until the fracture of the bottom face sheet.

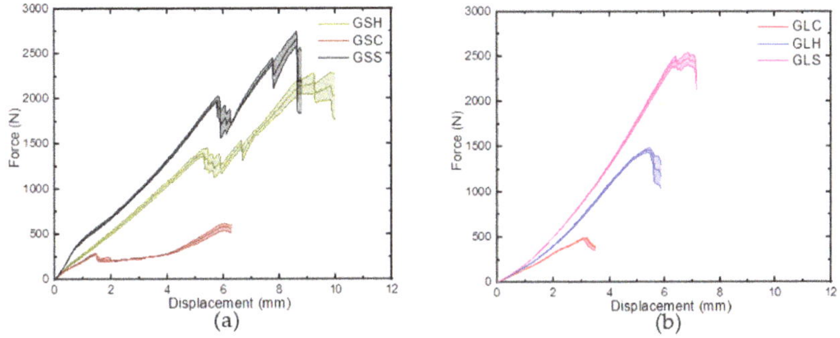

Figure 8. Load–displacement curves: (**a**) GS, (**b**) GL.

For the GL samples (Figure 8b), a similar quasi-brittle behaviour was observed as the load rapidly increased until the fracture of the fibres at 2.48 kN, 1.49 kN, and 0.48 kN for the GLs, GLh, and GLc specimens, respectively. The limited surface area of contact between the impacted material and the conical indenter, observed for GS samples and which resulted in smaller loads, might also have a similar effect on the laminate samples. It is noteworthy that while the load–displacement curve for the GS specimens considers the contribution of the core and bottom face sheets, the corresponding curve for the laminates was similar to the portion of the GS samples that related to the top face sheets with comparable thickness.

3.2. Energy-Absorption Properties

The total energy absorbed by all samples with the conical indenter was generally lower when compared to other indenter geometries (Figure 9). The GLC absorbed 13.3% and 28.0% of the energy of GLS and GLH before fracture, respectively. Similarly, GSC absorbed 14.1% and 17.7% of the energy of GSS and GSH before fracture, respectively. It is worth noting that the drops noticed in the top face sheets in the force-displacement diagrams of the GS samples were small, and thus there was no noticeable decline in E_a for the specimens.

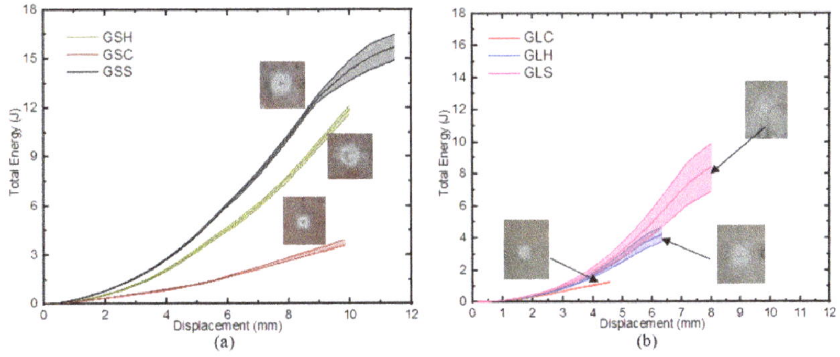

Figure 9. Total energy analysis: (**a**) GS, (**b**) GL.

A comparison of the energy-absorption capabilities per unit thickness of the GS and GL samples revealed that the former (GSS and GSH) had the highest potential for energy-absorption, while the samples indented by the conical indenter (GS_c and GL_c) had lower energy-absorption properties. The SEA of the GL and GS samples impacted with the square and hemispherical indenters differed significantly, by 19.2% and 43.3%, respectively (Figure 10). This is attributed to the presence of the core and the related increase in the sample thickness. However, for the samples tested under conical indentation, the energy-absorption curves showed an insignificant difference (3.5%) between the GS and GL samples. This was solely caused by the shape of the indenter, and thus limited the contributory effects of the core and through-thickness on the energy-absorption properties of the samples under such regimes.

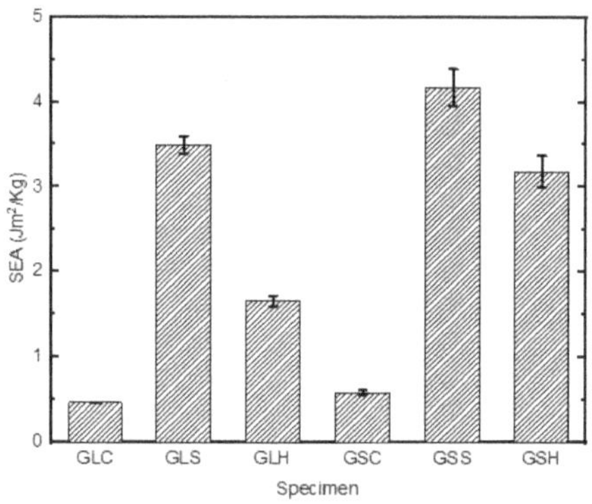

Figure 10. Specific energy absorption for different samples.

3.3. Acoustic Emission

3.3.1. Feature Analysis

The normalised force-time graphs of the samples under various indenter geometry with the corresponding AE signals are presented in Figure 11. The aim was to monitor the variation in AE features in order to identify failure points (drop in force) with respect to time. For the laminate samples (Figure 12a–c), crack initiation with significant plastic deformation was observed until failure, with a drop in the load curve at around 150 s, 180 s, and 100 s for GLH, GLS, and GLC, respectively. The increase in the amplitude of AE signals was attributed to crack propagation up to the critical load, beyond which the AE inputs could be regarded as noise from the experimental setup [38,39]. Similar observations were made for the GS samples (Figure 12d–f), with an increase in the amplitude of AE signals as the force progressed through two notable peaks corresponding to the top and bottom face sheet failures.

The failure of the top and bottom face sheets was characterised by higher amplitude signals (>50 dB) observed at around 150 s and 300 s for the GSH and GSS specimens, respectively. However, for the GSc specimen, higher amplitudes due to face sheet perforation occurred earlier (50 s and 150 s), which was attributed to the shape of the conical indenter, leading to lower damage resistance of the samples. As the conical indenter penetrated the sample, the contact area with the sample progressively increased from the tip to the base, resulting in the expansion of the damaged area and leading to the failure of more fibres, recorded as high amplitude signals over a longer time when compared to other GS samples. A similar indenter-induced damage recognition with AE was reported in [17,25]. It should be noted that in this study, high amplitude hits recorded prior to the onset of damage were assumed to be signals generated by the mechanical setup and not indicative of the damage mechanism in the samples.

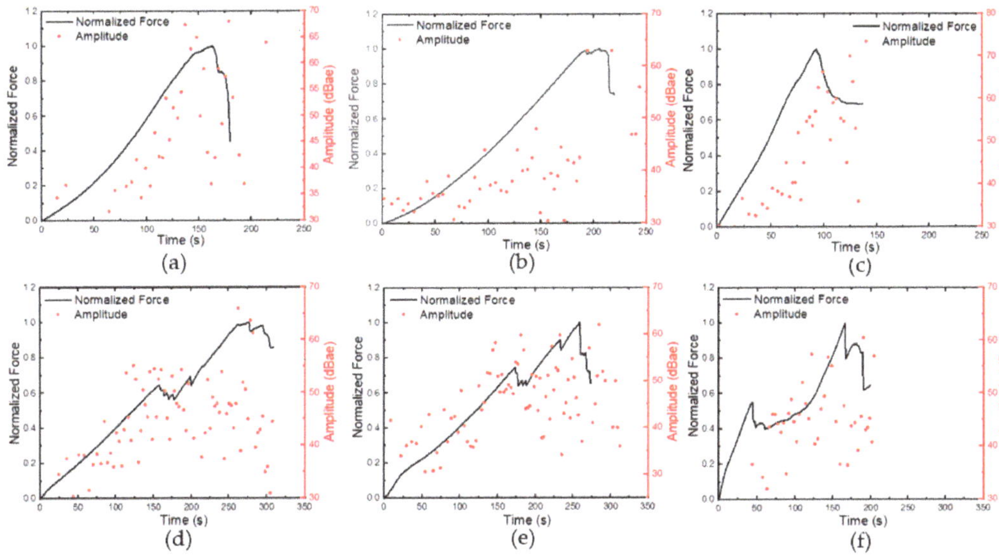

Figure 11. Average force-time curve with AE amplitude: (**a**) GLH, (**b**) GLS, (**c**) GLC, (**d**) GSH, (**e**) GSS, (**f**) GSC.

It was proven that the AE cumulative count behaviour provided a significant role in assessing the damage mechanism and failure characteristics of composite materials as it allows a substantial classification of distinct zones of damage under quasi-static loading conditions [40]. Therefore, to further clarify the damage mechanism corroborated with AE features, it was necessary to analyse the AE counts and duration features. A change in the gradient of the cumulative counts slope indicated a transition in the load-bearing capacity of the sample, reflecting of the presence of a damage sequence, which resulted in a corresponding increase in the duration hits [38]. As can be seen from Figure 12a–c, three distinct regions were identified for the laminates, while 5 were established regions for the sandwich samples.

For the laminate specimens, region I was identified as the initiation of damage at the microscopic stage (matrix cracking), region II represents the initiation of macroscopic damage such as fibre/matrix debonding, and finally region III indicates the start of damage propagation until failure of the structure (fibre breakage/rupture) [39]. On the other hand, two additional damage processes could be identified (regions IV and V) for samples, and the earlier damage points discussed were similar to those of the face sheet of the GS samples [17]. Region IV represents the core/face sheet debonding and delamination, while region V indicates the fracture of the bottom face sheets. It could be seen that the duration of region II (in Figure 12a–c) for the laminates decreased with a reduction in the contact surface area of the indenter, leading to a faster penetration and subsequent macroscopic damage.

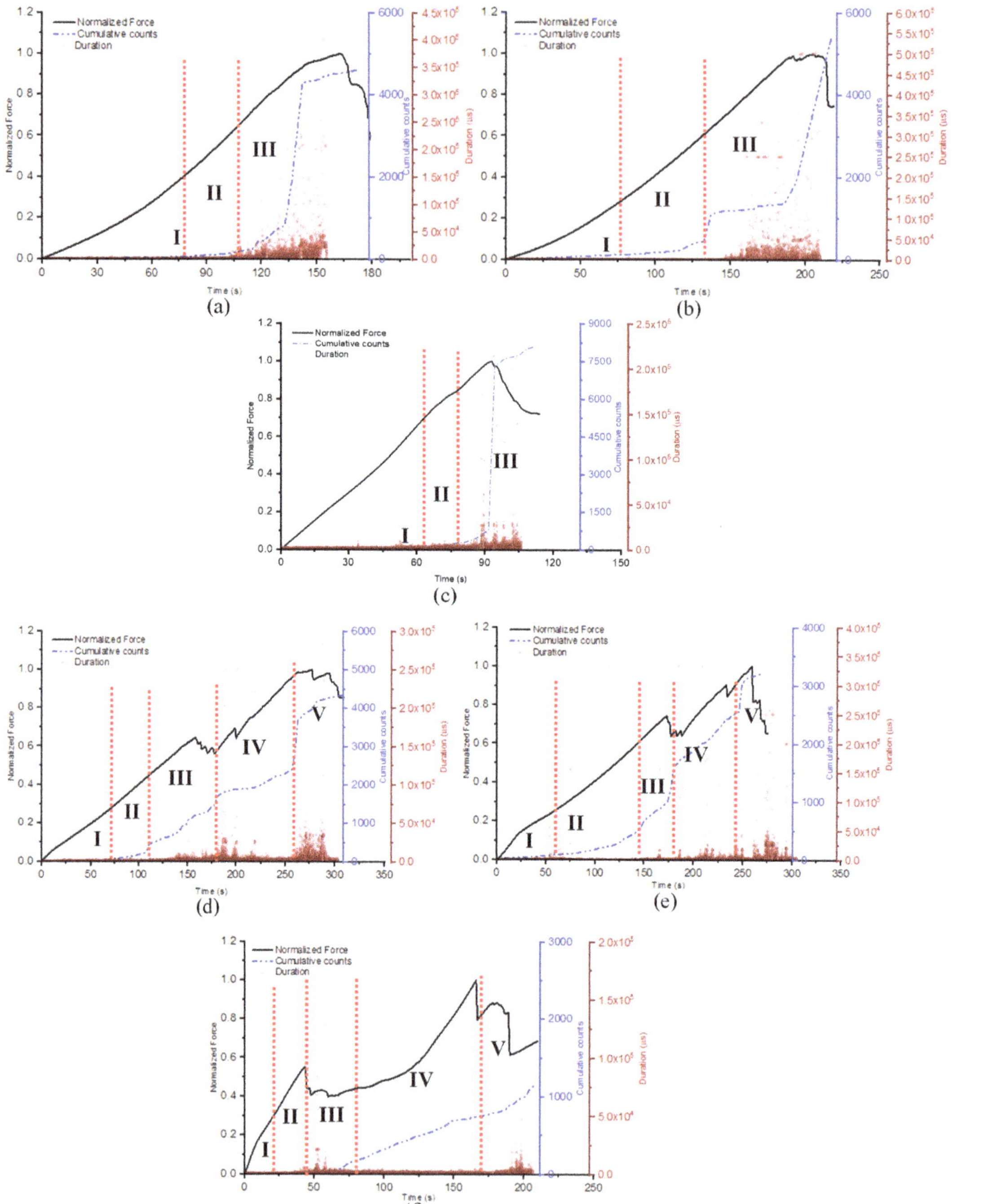

Figure 12. Normalised force-time curves with AE cumulative counts and duration: (**a**) GLH, (**b**) GLS, (**c**) GLC, (**d**) GSH, (**e**) GSS, (**f**) GSC.

Conversely, region IV (in Figure 12d–f) for the sandwich specimens experienced a much gentler slope of the AE cumulative counts and steady duration hits with an amplitude below 50 dB, indicative of the limited damage resistance offered by the core to the conical indenter up to bottom face sheet perforation. This could be attributed to the low stiffness/thickness ratio of the foam core, thereby resulting in the bottom face sheet primarily responsible for damage resistance of the structure.

3.3.2. Pattern-Recognition Analysis

The PCC for the selected AE features (defined in Section 2.3.2) for 6 specimens is shown in Figure 13. Among all features, the peak frequency, which is expected to be affected by attenuation, was selected for cluster analysis, together with the amplitude feature with the lowest time-domain PCC. The results for the PCC for the GS and GL samples are provided in Figure 13. As shown in Figure 14, the k-means++ clustering analysis separates the AE data into 3 and 4 clusters, representing the damage modes for the GS and GL specimens, respectively. The cluster validity evaluation scores for different specimens are presented in Table 3.

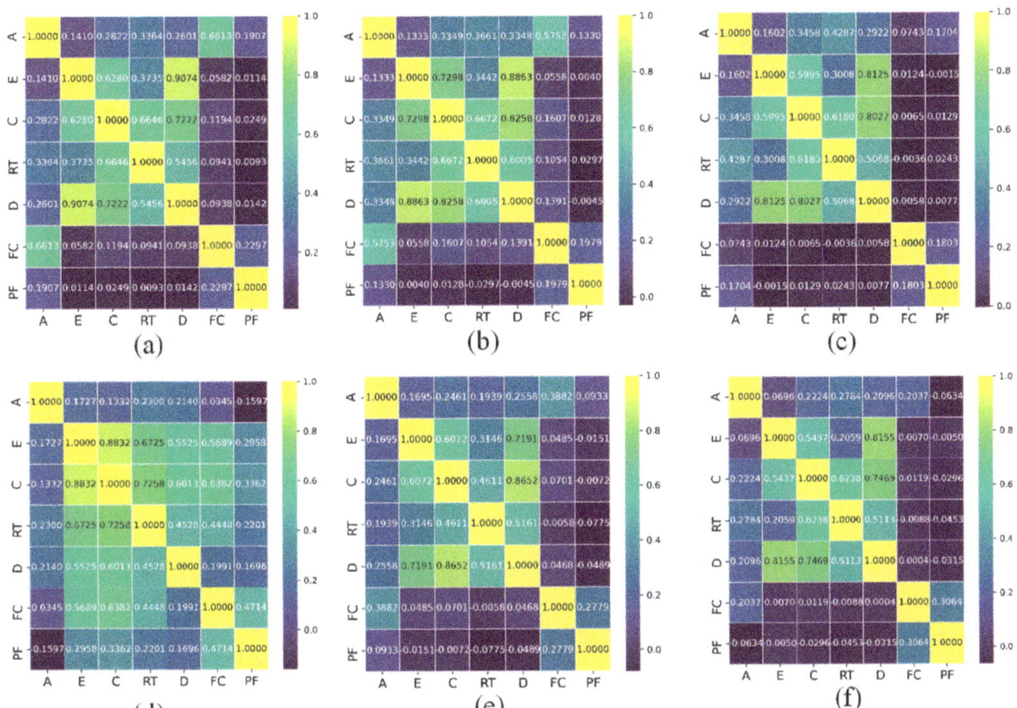

Figure 13. PCC between AE signals for different specimens: (**a**) GLH, (**b**) GLC, (**c**) GLS, (**d**) GSH, (**e**) GSC, (**f**) GSS (PF—peak frequency, A—amplitude, D—duration, C—counts, RT—rise time, E—energy, and FC—frequency centroid).

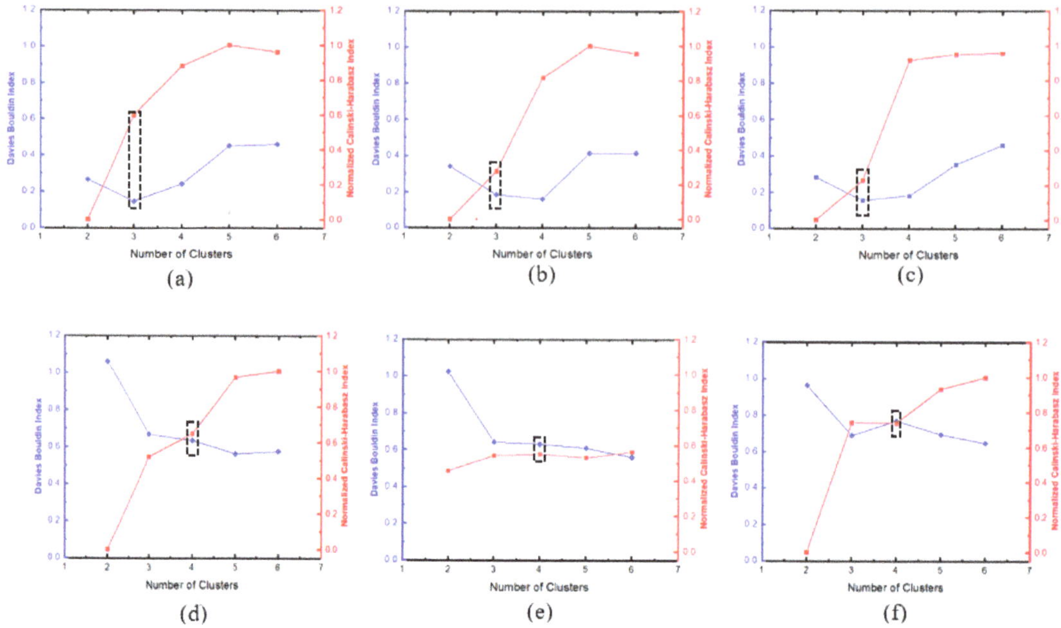

Figure 14. Pearson's correlation; (**a**) GLH, (**b**) GLC, (**c**) GLS, (**d**) GSH, (**e**) GSC, (**f**) GSS.

Table 3. PCC for various samples.

Specimen	PCC	
	DBI	**CHI**
GLH	0.147	0.595
GLC	0.185	0.277
GLS	0.156	0.230
GSH	0.635	0.649
GSC	0.631	0.550
GSS	0.770	0.739

The two-stage clusterisation approach proposed in [39] was adopted to determine the damage morphology of the samples. The first stage for the GS specimen was conducted with AE features divided into two broad classes. As discussed earlier, the damage mechanism in the GL samples was similar to that of the face sheets of the GS specimens and, thus, the additional cluster could be attributed to the damage in the core of the GS specimen via direct elimination method. This was also corroborated by the similarities in the AE events described in Figures 11 and 12. In Figure 15, cluster 0 represents the low-frequency features (<60 kHz), with a variation in amplitude (up to 60 dBae) indicative of a damage phenomenon, while cluster 1 represents features with frequencies above 75 kHz and up to 85 dBae. The frequency range for core damage can be delineated according to the previous literature [39,41].

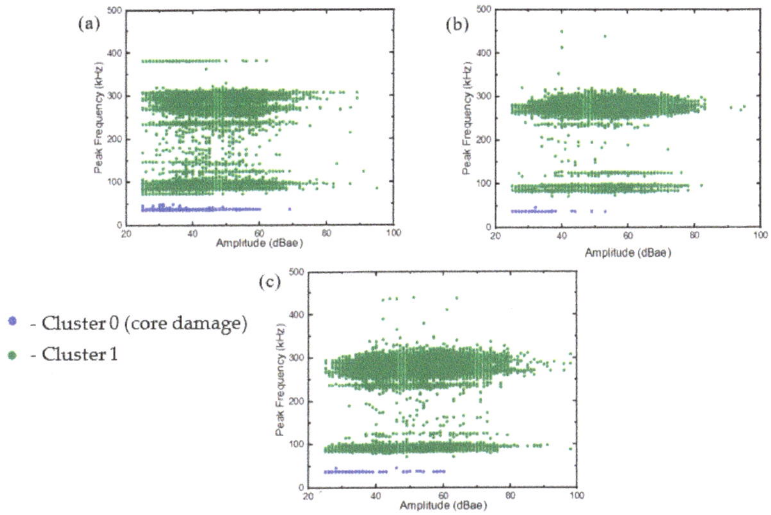

Figure 15. Stage 1 clustering: (**a**) GSH, (**b**) GSC, (**c**) GSS.

Stage 2 of the clusterisation involved the processing of all the remaining AE features in the GS and GL samples. Three distinct clusters were observed (see Figure 16), and the corresponding damage sequence was attributed to the laminates and face sheets of the GS samples. The results revealed that the frequency content attributed to matrix cracking was recorded within the range of 60–140 kHz, with most hits in the range between 90 and 110 kHz, similar to the results reported in [16,28]. This range could be observed both in GL and GS samples. Similarly, further phenomenological damage, such as fibre/matrix debonding and fibre breakage, was also obtained from the analysis. Two higher frequency damage modes identified in the tested specimens were cluster hits at 180–315 kHz and 360–453 kHz, with a concentration of hits at 222–315 kHz and 453 kHz, corresponding to fibre/matrix debonding and fibre breakage, respectively.

These results are similar to the values reported in other studies [42]. In general, the fourth damage cluster observed in the GS samples and the three clusters in the GL specimens were corroborated by the PCC analysis of the AE hits for the GS samples discussed above.

3.4. *Damage and Failure Morphology*

3.4.1. X-ray μ-CT Results

After applying the AE clusterisation technique, the identified damage morphology was validated using X-ray μ-CT and SEM images. For the μ-CT investigation, three samples per configuration were examined, and parameters such as area, perimeter, and depth were computed and presented. It has been established that understanding BVID is crucial, since out-of-plane energy is dissipated through the internal damage mechanism (such as delamination, debonding, and matrix crack) of the structures [43–45]. The 3D, top, and cross-sectional views of the undamaged, GSS, GSH, and GSC samples are shown in Figure 17I–IV. All images were taken at the impact region where maximum damage occurred, and various damage mechanisms were revealed for the different indenter configurations.

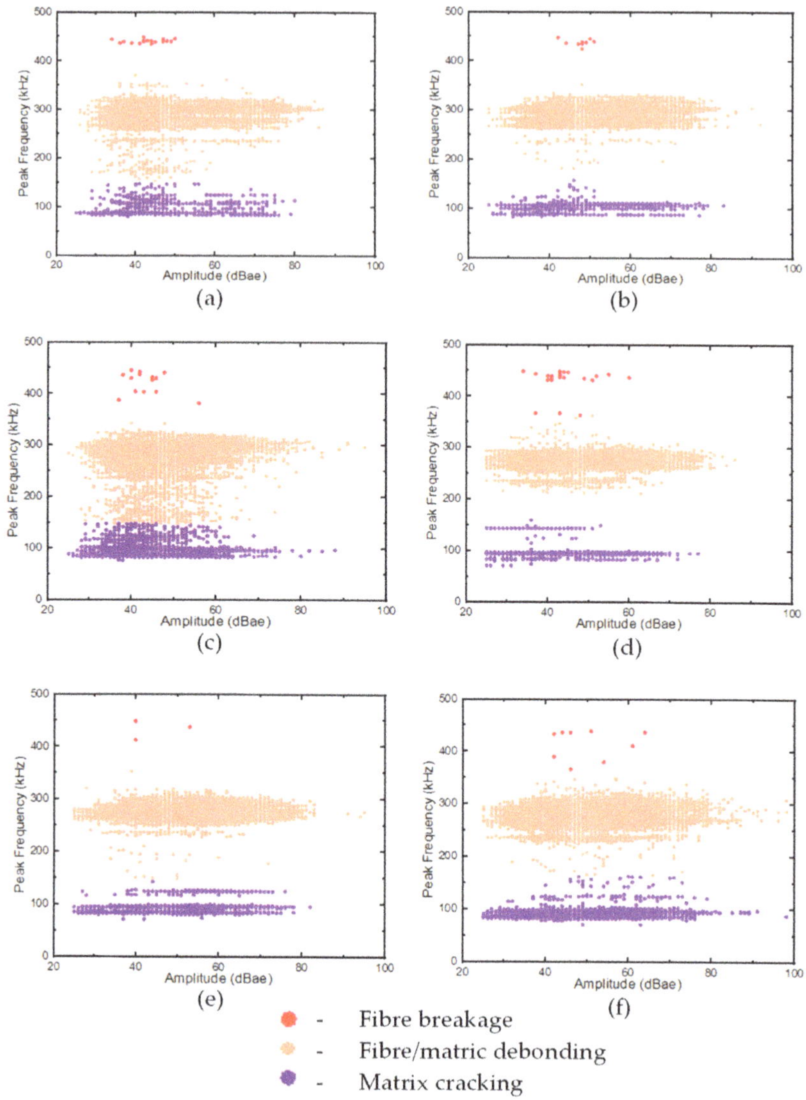

Figure 16. Stage 2 clustering (**a**) GLH, (**b**) GLC, (**c**) GLS, (**d**) GSH, (**e**) GSC, (**f**) GSS.

The damage evolution was more pronounced in samples indented with the hemispherical and square impactors than with the conical indenter, as evidenced by a larger damaged area. Apparently (Figure 17), the continuous loading of the sample during the QSI test caused the foam core to experience severe central crushing for the GSS and GSH samples, while all samples experienced core side shearing. For samples with a higher contact area with the indenter, the predominant damage mechanism included core crushing, delamination of the bottom face sheet, and the abrupt fibre fracture of the top face sheets. Interestingly, this fracture was identical to the indenter configuration in the GSS samples. This could also be attributed to the brittleness of the reinforcements and the high friction as the indenter moved through the specimen. Furthermore, the extent of the fibre breakage in the bottom face sheet was less severe for the GSS and GSH samples when compared to the GSC ones. This is attributed to the shape of the conical indenter; moving through the sample with a lower contact area resulted in greater penetration of the indenter as well as limited frictional resistance from the face sheet and the core. Further, a post mortem analysis of the GL samples (Figure 18) revealed damage characteristics similar to those of the GS samples impacted with the square and hemispherical indenters. Interlaminar delamination and fibre breakage were identified as the dominant damage mechanisms, while for the conical indenter damage was more localised within a smaller area, which could be attributed to the geometry of the indenter and the brittle nature of the reinforcement. It is worth noting that the damage morphology of the laminates is understandably similar to that of the top face sheet of the sandwich samples for all the indentation configurations. These were all consistent with the results obtained from the AE signal clusterisation.

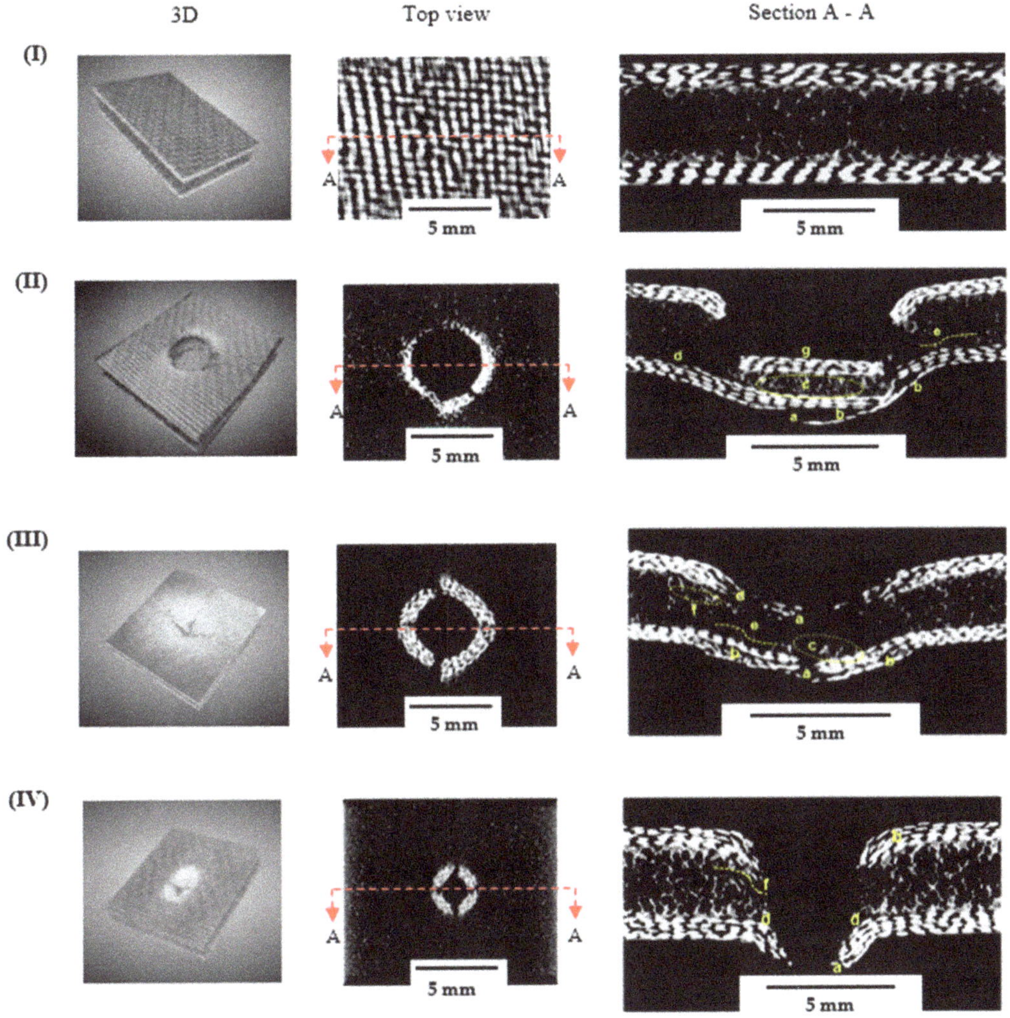

Figure 17. X-ray μ-CT analysis for GS samples: (**I**) undamaged (**II**) GSS, (**III**) GSH, (**IV**) GSC note; a—fibre breakage, b—delamination, c—core crushing, d—core/face sheet debonding, e—core fracture crack, f—core compression, g—compressed face sheets.

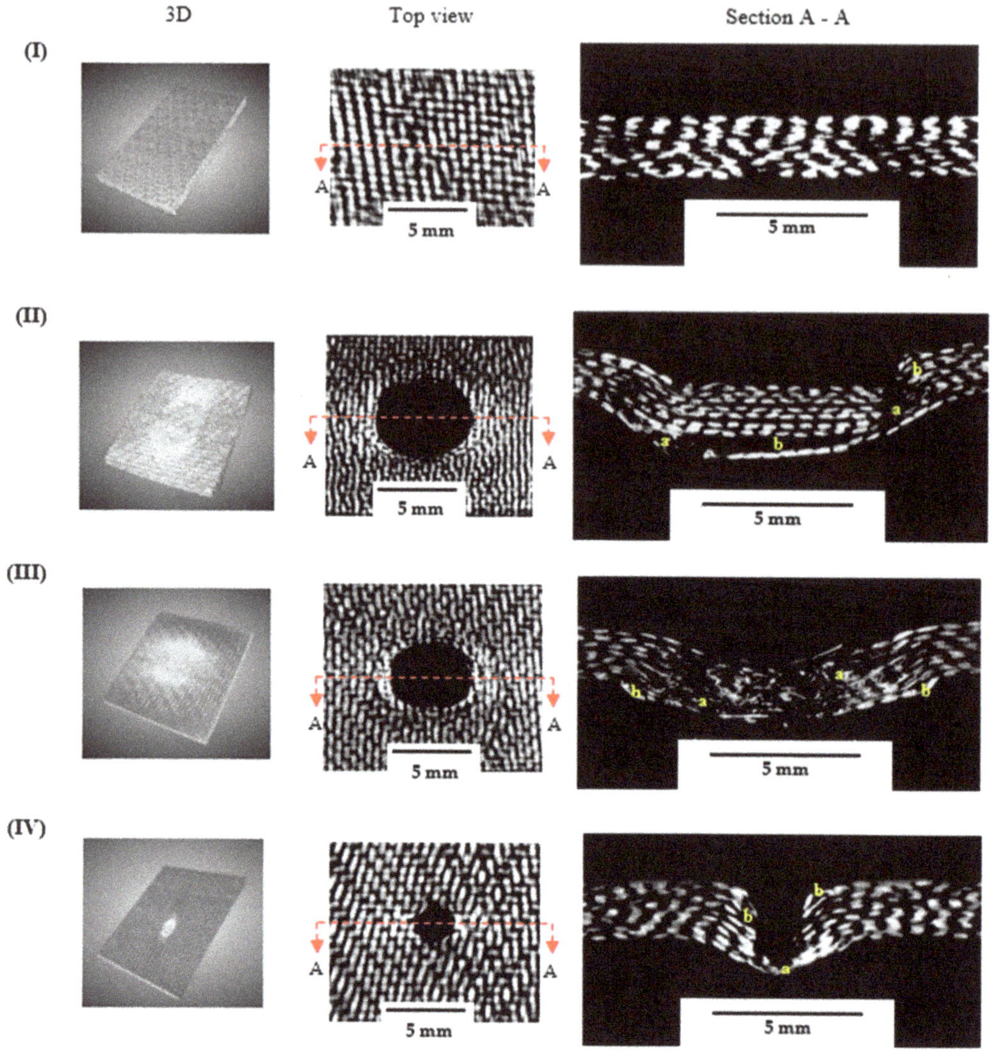

Figure 18. Post mortem μ-CT analysis for GL samples: (**I**) undamaged, (**II**) GLS, (**III**) GLH, (**IV**) GLC. Note: a—fibre breakage, b—delamination.

Additionally, the effect of the indenter shape on the energy could be evaluated from the area, perimeter, and damage depth per thickness (d/t) of damage in the samples. It was observed that the damaged area and perimeter for the square and hemispherical indenters were higher than those for the conical indenters. However, d/t was similar for almost all samples, except those indented with the square indenter, which was considerably higher, at 1.8 and 2.02 for GSS and GLS, respectively. This could be due to the shape of the indenter, which prevented easy penetration, thereby increasing the friction between the indenter and the constituents of the sample. Added to this phenomenon is the brittle nature of the fibres, which caused abrupt damage at the top face sheet and laminate (Figures 17II and 18II). A summary of the damage parameters measured from the post mortem X-ray CT scan of the samples loaded with various indenter geometries is listed in Table 4.

Table 4. Summary of damage parameters for GS and GL samples.

Specimen	Peak Load (N)	Max Energy (J)	Area (mm^2)	Perimeter (mm)	d/t
GSH	2259.38 ± 71.88	12.07 ± 0.33	150.96 ± 16.94	47.09 ± 2.03	1.49 ± 0.02
GSC	589.60 ± 22.91	2.14 ± 0.10	37.61 ± 1.58	23.77 ± 0.15	1.34 ± 0.02
GSS	2774.60 ± 59.17	15.20 ± 0.71	119.59 ± 9.03	48.42 ± 1.12	1.80 ± 0.05
GLH	1488.40 ± 24.98	4.58 ± 0.19	48.04 ± 5.03	26.55 ± 1.17	1.48 ± 0.13
GLC	499.93 ± 7.58	1.28 ± 0.02	10.23 ± 1.46	13.67 ± 0.39	1.5 ± 0.09
GLS	2259.83 ± 71.88	9.70 ± 0.27	76.70 ± 9.44	32.59 ± 1.74	2.02 ± 0.08

3.4.2. SEM Analysis

To obtain a deeper understanding of the damage morphology at the micro-level, SEM images of damaged samples taken at 250× magnification are presented in Figure 19. Higher magnification revealed that the glass fibre had a section of bare surface (Figure 19), which is indicative of the lack of adhesion between the matrix and the fibres. Furthermore, sections of the images showed a wave-like damage morphology, which indicates the presence of the matrix cracking damage mode. Thus, at the microscale, the presence of these damage modes (matrix cracking, fibre breakage, and fibre breakage with pull-out) corroborated the morphological damage phenomenon earlier observed in the μ-CT images; a similar trend was also reported by [46]. Having demonstrated the effectiveness of this methodology for exploring the damage morphology of the FRPs, it is suggested that this approach could be expanded to elucidating the energy absorption and damage sequence of FRPs subjected to extreme environmental conditions, such as seawater and Arctic temperatures.

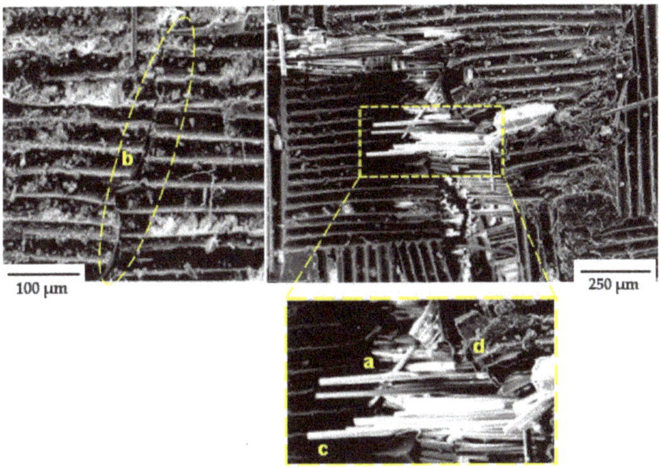

Figure 19. SEM analysis of GS face sheet and GL: a—fibre breakage, b—matrix crack, c—fibre pull-out, d—fibre/matrix interface debonding.

4. Conclusions

The energy-absorption properties and the damage morphology of GFRP laminates and sandwich structures with the GFRP face sheet with a PVC foam core when subjected to indenters with different shapes were investigated. The experimental results obtained from the QSI tests, as well as AE monitoring, produced the following conclusions.

Under QSI conditions, the contact area with the indenter played a crucial role in the damage resistance of the GS and GL samples. For the former, the conical indenter with the surface contact exhibited lower loads at the top face sheet (28.3% and 26.3% of values for GSH and GSS samples, respectively) and complete penetration (20.1% and 14.1% of the respective values for GSH and GSS). Similar results were obtained for the GL samples, with lower loads at fracture of 2.5 kN, 1.7 kN, and 0.48 kN for the GLS, GLH, and GLC specimens, respectively. Relatedly, lower energy-absorption capabilities were presented by GSC samples (14.1% and 17.8% of the energy of GSS and GSH, respectively) and GL_c (13.25% and 28.03% of the energy of GL_s and GL_h before fracture, respectively).

The unsupervised pattern recognition analysis of the AE events showed that the GS had four clusters, while the GL had three clusters. These clusters were linked to the damage mechanisms identified through post mortem X-ray μ-CT and SEM studies of the samples, namely, matrix cracking, fibre/matrix debonding, core shearing/crushing, and fibre breakage. The damage-characterisation methodology proposed in this study, therefore, proved that the AE technique could underpin the understanding of failure modes in composite structures by identifying the main damage mechanisms, including those experienced by the cores of sandwich materials. The data obtained could be useful for the optimisation of the materials, while the applied methodology could be expanded to investigate damage in FRP structures subjected to varying environmental factors. Furthermore, as a future study, the effects of the varied thickness of GS and the number of layers of GL on the failure modes could be investigated with the developed AE methodology.

Author Contributions: Conceptualisation, N.O.-u., V.V.S. and E.D.; methodology, N.O.-u., V.V.S. and E.D.; software, N.O.-u. and A.G.U.; formal analysis, N.O.-u. and A.G.U.; resources, N.O.-u., K.P.B. and E.D.; data curation, N.O.-u. and A.G.U.; writing—original draft preparation, N.O.-u. and A.G.U.; writing—review and editing, V.V.S., K.P.B. and E.D.; visualisation, N.O.-u. and A.G.U.; supervision, V.V.S. and E.D. All authors have read and agreed to the published version of the manuscript.

Funding: This research was funded by the Nigerian Air Force, grant number OPS/1282DTG27145AJUL21. KPB acknowledges support from the Royal Society, grant number RGS\R1\221368.

Institutional Review Board Statement: Not applicable.

Data Availability Statement: The data presented in this study are available on request from the corresponding author.

Conflicts of Interest: The authors declare no conflict of interest.

References

1. Balıkoğlu, F.; Arslan, N.; Demircioğlu, T.K.; İnal, O.; İren, M.; Ataş, A. Improving Four-Point Bending Performance of Marine Composite Sandwich Beams by Core Modification. *J. Compos. Mater.* **2020**, *54*, 1049–1066. [CrossRef]
2. Altin Karataş, M.; Gökkaya, H. A Review on Machinability of Carbon Fiber Reinforced Polymer (CFRP) and Glass Fiber Reinforced Polymer (GFRP) Composite Materials. *Def. Technol.* **2018**, *14*, 318–326. [CrossRef]
3. Le Duigou, A.; Davies, P.; Baley, C. Seawater Ageing of Flax/Poly(Lactic Acid) Biocomposites. *Polym. Degrad. Stab.* **2009**, *94*, 1151–1162. [CrossRef]
4. Kausar, A.; Ahmad, I.; Rakha, S.A.; Eisa, M.H. State-Of-The-Art of Sandwich Composite Structures: Manufacturing-to-High Performance Applications. *J. Compos. Sci.* **2023**, *7*, 102. [CrossRef]
5. Osa-uwagboe, N.; Silberschimdt, V.V.; Aremi, A.; Demirci, E. Mechanical Behaviour of Fabric-Reinforced Plastic Sandwich Structures: A State-of-the-Art Review. *J. Sandw. Struct. Mater.* **2023**, *25*, 591–622. [CrossRef]
6. Fan, Y.; Wang, Y. The Effect of Negative Poisson's Ratio on the Low-Velocity Impact Response of an Auxetic Nanocomposite Laminate Beam. *Int. J. Mech. Mater. Des.* **2021**, *17*, 153–169. [CrossRef]
7. Sutherland, L.S.; Guedes Soares, C. The Use of Quasi-Static Testing to Obtain the Low-Velocity Impact Damage Resistance of Marine GRP Laminates. *Compos. B Eng.* **2012**, *43*, 1459–1467. [CrossRef]
8. Wagih, A.; Maimí, P.; Blanco, N.; Costa, J. A Quasi-Static Indentation Test to Elucidate the Sequence of Damage Events in Low Velocity Impacts on Composite Laminates. *Compos. Part A Appl. Sci. Manuf.* **2016**, *82*, 180–189. [CrossRef]
9. Zniker, H.; Ouaki, B.; Bouzakraoui, S.; EbnTouhami, M.; Mezouara, H. Energy Absorption and Damage Characterization of GFRP Laminated and PVC-Foam Sandwich Composites under Repeated Impacts with Reduced Energies and Quasi-Static Indentation. *Case Stud. Constr. Mater.* **2022**, *16*, e00844. [CrossRef]
10. Raju, K.S.; Smith, B.L.; Tomblin, J.S.; Liew, K.H.; Guarddon, J.C. Impact Damage Resistance and Tolerance of Honeycomb Core Sandwich Panels. *J. Compos. Mater.* **2008**, *42*, 385–412. [CrossRef]
11. Daniel, I.M.; Abot, J.L.; Schubel, P.M.; Luo, J.J. Response and Damage Tolerance of Composite Sandwich Structures under Low Velocity Impact. *Exp. Mech.* **2012**, *52*, 37–47. [CrossRef]
12. Schubel, P.M.; Luo, J.J.; Daniel, I.M. Impact and Post Impact Behavior of Composite Sandwich Panels. *Compos. Part A Appl. Sci. Manuf.* **2007**, *38*, 1051–1057. [CrossRef]

13. Zhu, F.; Lu, G.; Ruan, D.; Wang, Z. Plastic Deformation, Failure and Energy Absorption of Sandwich Structures with Metallic Cellular Cores. *Int. J. Prot. Struct.* **2010**, *1*, 507–541. [CrossRef]
14. Liu, P.F.; Yang, J.; Peng, X.Q. Delamination Analysis of Carbon Fiber Composites under Hygrothermal Environment Using Acoustic Emission. *J. Compos. Mater.* **2017**, *51*, 1557–1571. [CrossRef]
15. Zhou, W.; Qin, R.; Han, K.N.; Wei, Z.Y.; Ma, L.H. Progressive Damage Visualization and Tensile Failure Analysis of Three-Dimensional Braided Composites by Acoustic Emission and Micro-CT. *Polym. Test.* **2021**, *93*, 106881. [CrossRef]
16. Verma, L.; Andrew, J.J.; Sivakumar, S.M.; Balaganesan, G.; Vedantam, S.; Dhakal, H.N. Evaluation of Quasi-Static Indentation Response of Superelastic Shape Memory Alloy Embedded GFRP Laminates Using AE Monitoring. *Polym. Test.* **2021**, *93*, 106942. [CrossRef]
17. Dikshit, V.; Nagalingam, A.P.; Goh, G.D.; Agarwala, S.; Yeong, W.Y.; Wei, J. Quasi-Static Indentation Analysis on Three-Dimensional Printed Continuous-Fiber Sandwich Composites. *J. Sandw. Struct. Mater.* **2021**, *23*, 385–404. [CrossRef]
18. Hajikhani, M.; Ahmadi, M.; Farjpour, M.; Oskouei, A.R.; Sharifi, A. Strain Energy Release Rate Assessment in Mode i Delamination of Foam Core Sandwich Composites by Acoustic Emission. *J. Compos. Mater.* **2011**, *45*, 2271–2277. [CrossRef]
19. Ben Ammar, I.; Karra, C.; El Mahi, A.; El Guerjouma, R.; Haddar, M. Mechanical Behavior and Acoustic Emission Technique for Detecting Damage in Sandwich Structures. *Appl. Acoust.* **2014**, *86*, 106–117. [CrossRef]
20. Guo, Y.; Zhu, S.; Chen, Y.; Li, D. Analysis and Identification of the Mechanism of Damage and Fracture of High-Filledwood Fiber/Recycled High-Density Polyethylene Composites. *Polymers* **2019**, *11*, 170. [CrossRef]
21. Tang, J.; Soua, S.; Mares, C.; Gan, T.H. A Pattern Recognition Approach to Acoustic Emission Data Originating from Fatigue of Wind Turbine Blades. *Sensors* **2017**, *17*, 2507. [CrossRef] [PubMed]
22. Šofer, M.; Cienciala, J.; Fusek, M.; PavlíCek, P.; Moravec, R. Damage Analysis of Composite Cfrp Tubes Using Acoustic Emission Monitoring and Pattern Recognition Approach. *Materials* **2021**, *14*, 786. [CrossRef] [PubMed]
23. Li, L.; Lomov, S.V.; Yan, X.; Carvelli, V. Cluster Analysis of Acoustic Emission Signals for 2D and 3D Woven Glass/Epoxy Composites. *Compos. Struct.* **2014**, *116*, 286–299. [CrossRef]
24. Eltaher, M.A.; Alsulami, R.; Wagih, A. On the Evolution of Energy Dissipation in Dispersed Composite Laminates under Out-of-Plane Loading. *Compos. B Eng.* **2021**, *216*, 108864. [CrossRef]
25. Dikshit, V.; Nagalingam, A.P.; Yap, Y.L.; Sing, S.L.; Yeong, W.Y.; Wei, J. Investigation of Quasi-Static Indentation Response of Inkjet Printed Sandwich Structures under Variousindenter Geometries. *Materials* **2017**, *10*, 290. [CrossRef]
26. EASYCell 75 Closed-Cell PVC Foam Specifications Datasheet. Available online: https://www.easycomposites.co.uk (accessed on 20 June 2023).
27. Reis, J.M.L.; Coelho, J.L.V.; Monteiro, A.H.; Da Costa Mattos, H.S. Tensile Behavior of Glass/Epoxy Laminates at Varying Strain Rates and Temperatures. *Compos. B Eng.* **2012**, *43*, 2041–2046. [CrossRef]
28. Jefferson Andrew, J.; Arumugam, V.; Ramesh, C.; Poorani, S.; Santulli, C. Quasi- Static Indentation Properties of Damaged Glass/Epoxy Composite Laminates Repaired by the Application of Intra-Ply Hybrid Patches. *Polym. Test.* **2017**, *61*, 132–145. [CrossRef]
29. Geren, N.; Acer, D.C.; Uzay, C.; Bayramoglu, M. The Effect of Boron Carbide Additive on the Low-Velocity Impact Properties of Low-Density Foam Core Composite Sandwich Structures. *Polym. Compos.* **2021**, *42*, 2037–2049. [CrossRef]
30. ASTM International. *D6264/D6264M-17*; Standard Test Method for Measuring the Damage Resistance of a Fiber-Reinforced Polymer-Matrix Composite to a Concentrated Quasi-Static Indentation Force. ASTM International: West Conshohocken, PA, USA, 2018; Volume 98, pp. 1–12. [CrossRef]
31. Cai, J.; Luo, J.; Wang, S.; Yang, S. Feature Selection in Machine Learning: A New Perspective. *Neurocomputing* **2018**, *300*, 70–79. [CrossRef]
32. Li, J.; Cheng, K.; Wang, S.; Morstatter, F.; Trevino, R.P.; Tang, J.; Liu, H. Feature Selection: A Data Perspective. *ACM Comput. Surv.* **2017**, *50*, 1–45. [CrossRef]
33. Chandrashekar, G.; Sahin, F. A Survey on Feature Selection Methods. *Comput. Electr. Eng.* **2014**, *40*, 16–28. [CrossRef]
34. Joseph, R.; Bhuiyan, Y.; Giurgiutiu, V. Acoustic Emission from Vibration of Cracked Sheet-Metal Samples. *Eng. Fract. Mech.* **2019**, *217*, 106544. [CrossRef]
35. Liu, Y.; Zhang, L.; Li, Z.; Chen, Z.; Huang, K.; Guo, L. Investigation on Damage Evolution of Open-Hole Plain Woven Composites under Tensile Load by Acoustic Emission Signal Analysis. *Compos. Struct.* **2022**, *305*, 116481. [CrossRef]
36. Saeedifar, M.; Zarouchas, D. Damage Characterization of Laminated Composites Using Acoustic Emission: A Review. *Compos. B Eng.* **2020**, *195*, 108039. [CrossRef]
37. Wisner, B.; Mazur, K.; Perumal, V.; Baxevanakis, K.P.; An, L.; Feng, G.; Kontsos, A. Acoustic Emission Signal Processing Framework to Identify Fracture in Aluminum Alloys. *Eng. Fract. Mech.* **2019**, *210*, 367–380. [CrossRef]
38. Wu, Y.; Pastor, M.L.; Perrin, M.; Casari, P.; Gong, X. A New Methodology to Predict Moisture Effects on Mechanical Behaviors of GFRP-BALSA Sandwich by Acoustic Emission and Infrared Thermography. *Compos. Struct.* **2022**, *287*, 115342. [CrossRef]
39. Wu, Y.; Pastor, M.-L.; Perrin, M.; Casari, P.; Gong, X. Characterisation of Damage Mechanisms of GFRP-Balsa Sandwich under 4-Point Bending Based on Two-Step Clustering Process in Acoustic Emission Analysis. *Compos. B Eng.* **2023**, *260*, 110774. [CrossRef]
40. Jefferson Andrew, J.; Arumugam, V.; Bull, D.J.; Dhakal, H.N. Residual Strength and Damage Characterization of Repaired Glass/Epoxy Composite Laminates Using A.E. and D.I.C. *Compos. Struct.* **2016**, *152*, 124–139. [CrossRef]

41. Pashmforoush, F.; Khamedi, R.; Fotouhi, M.; Hajikhani, M.; Ahmadi, M. Damage Classification of Sandwich Composites Using Acoustic Emission Technique and K-Means Genetic Algorithm. *J. Nondestr. Eval.* **2014**, *33*, 481–492. [CrossRef]
42. Andrew, J.J.; Arumugam, V.; Ramesh, C. Acoustic Emission Characterization of Local Bending Behavior for Adhesively Bonded Hybrid External Patch Repaired Glass/Epoxy Composite Laminates. *Struct. Health Monit.* **2019**, *18*, 739–756. [CrossRef]
43. Elamin, M.; Li, B.; Tan, K.T. Impact Damage of Composite Sandwich Structures in Arctic Condition. *Compos. Struct.* **2018**, *192*, 422–433. [CrossRef]
44. Aktay, L.; Johnson, A.F.; Holzapfel, M. Prediction of Impact Damage on Sandwich Composite Panels. *Comput. Mater. Sci.* **2005**, *32*, 252–260. [CrossRef]
45. Adeniran, O.; Osa-Uwagboe, N.; Cong, W.; Ramoni, M. Fabrication Temperature-Related Porosity Effects on the Mechanical Properties of Additively Manufactured CFRP Composites. *J. Compos. Sci.* **2023**, *7*, 12. [CrossRef]
46. Gholami, M.; Afrasiab, H.; Baghestani, A.M.; Fathi, A. Mechanical and Failure Analysis of Thick Composites under Hygrothermal Conditions by a Novel Coupled Hygro-Thermo-Mechanical Multiscale Algorithm. *Compos. Sci. Technol.* **2022**, *230*, 109773. [CrossRef]

materials

MDPI

Article

Damage Detection in a Polymer Matrix Composite from 4D Displacement Field Measurements

Ana Mandić [1,2], Viktor Kosin [2,3], Clément Jailin [4], Zvonimir Tomičević [1,*], Benjamin Smaniotto [2] and François Hild [2,*]

1 Faculty of Mechanical Engineering and Naval Architecture, University of Zagreb, 10000 Zagreb, Croatia
2 Université Paris-Saclay, CentraleSupélec, ENS Paris-Saclay, CNRS, LMPS–Laboratoire de Mécanique Paris-Saclay, 91190 Gif-sur-Yvette, France
3 Institut für Angewandte Mathematik (IfAM), Leibniz Universität Hannover, 30167 Hannover, Germany
4 GE HealthCare, 78530 Buc, France; clement.jailin@ge.com
* Correspondence: zvonimir.tomicevic@fsb.hr (Z.T.); francois.hild@ens-paris-saclay.fr (F.H.)

Abstract: Standard Digital Volume Correlation (DVC) approaches enable quantitative analyses of specimen deformation to be performed by measuring displacement fields between discrete states. Such frameworks are thus limited by the number of scans (due to acquisition duration). Considering only one projection per loading step, Projection-based Digital Volume Correlation (P-DVC) allows 4D (i.e., space and time) full-field measurements to be carried out over entire loading histories. The sought displacement field is decomposed over a basis of separated variables, namely, temporal and spatial modes. In the present work, the spatial modes are constructed via scan-wise DVC, and only the temporal amplitudes are sought via P-DVC. The proposed method is applied to a glass fiber mat reinforced polymer specimen containing a machined notch, subjected to in situ cyclic tension and imaged via X-ray Computed Tomography. The P-DVC enhanced DVC method employed herein enables for the quantification of damage growth over the entire loading history up to failure.

Keywords: polymer matrix composite; projection based digital volume correlation; damage growth; correlation residuals

Citation: Mandić, A.; Kosin, V.; Jailin, C.; Tomičević, Z.; Smaniotto, B.; Hild, F. Damage Detection in a Polymer Matrix Composite from 4D Displacement Field Measurements. *Materials* **2023**, *16*, 6300. https://doi.org/10.3390/ma16186300

Academic Editors: Luís Miguel Pereira Durão and Nuno Calçada Loureiro

Received: 7 August 2023
Revised: 13 September 2023
Accepted: 15 September 2023
Published: 20 September 2023

1. Introduction

As it provides 3D images of scanned microstructures in a non-destructive way, X-ray Computed Tomography (XCT) has given a major impetus to the field of mechanics of materials [1,2]. With the recent developments of testing machines, 3D full-field displacement fields can be quantified in situ by coupling XCT and Digital Volume Correlation (DVC) into unique frameworks [3,4]. As DVC is based upon matching the gray levels between fully reconstructed volumes, the major limitations of such approach are its low temporal sampling and experiment duration. Each reconstructed volume requires from few minutes up to few hours of radiograph series to be acquired. This process governs the number of possible acquired scans. As the material response between two consecutive scans is not accessible, the quantification of time-dependent phenomena (e.g., stress-relaxation or crack propagation) is restricted. Furthermore, due to relaxation during in situ experiments, the reconstructed volumes may be impacted by motion artifacts. This issue can be avoided by starting the radiograph acquisition only after the measured force has stabilized. The aforementioned restrictions may be overcome by two different routes. The first route is to perform fast acquisitions, say in synchrotron facilities, with extremely bright X-ray beams using high-speed cameras. However, synchrotron imaging presents some drawbacks such as poor accessibility, high doses and the reconstruction quality that may degrade due to blur induced by vibrations [5,6]. Another route is to utilize the recently developed Projection-based Digital Volume Correlation (P-DVC), which aims to surpass the low temporal resolution attributed to classical (i.e., scan-wise) DVC approaches [7]. This method enables for the measurement of 4D (i.e., space and time) displacement fields from series of 2D radiographs acquired at different angles and load levels (instead of working with series of reconstructed 3D volumes). In such a way, huge gains in acquisition time and testing duration can be reached [4,8,9].

The ever-increasing application of fiber-reinforced polymers (FRPs) across various industrial branches is attributed to their advantageous properties such as high stiffness-to-weight ratios, which outperform conventional engineering materials [10]. Due to the occurrence of various damage mechanisms at different scales under mechanical loading, XCT is appealing in damage analyses of FRPs. The latter has been employed to quantify damage growth in FRPs under various loading conditions by, e.g., analyzing volumes [11–14]. However, it is worth noting that with this approach, only microstructural changes were evaluated, while it cannot provide access to the bulk kinematics, nor can it correlate it with microstructural changes. This limitation can be surpassed by DVC, which enables for 3D full-field displacement measurements [15,16]. Local approaches to DVC, which do not assume continuity of the displacement fields as they split the investigated Region of Interest (ROI) into smaller zones and register them independently, have been employed in the analysis of FRPs [17–19]. Conversely, global approaches based on, e.g., finite element (FE) discretizations were introduced more recently [20,21]. Such approaches are based on the assumption of continuity of displacement fields. A significant advantage of FE-DVC is the availability of correlation residuals. They correspond to the gray level differences between the reference volume and that of the deformed material corrected by the measured displacement field [4]. The conservation of gray levels is the underlying hypothesis of DVC [16,22,23]. Therefore, the correlation residuals can be used to check the quality of the registration. The initiation and growth of damage violates gray level conservation, thus the analysis of correlation residual fields (for converged displacement fields) reveal discontinuities corresponding to damaged zones [4,24,25]. Furthermore, FRPs are likely to undergo stress relaxation during tests [26,27]. As reconstructed volumes may last for up to hours, the specimen should be in a steady state to avoid poor image quality due to motion artifacts. Therefore, performing full-field displacement measurements in space and time is crucial especially for FRPs. The full insight into the material behavior is then provided in situ, and it is fully coupled with changes in the underlying microstructure.

In general, P-DVC relies on spacetime discretizations of the sought displacement fields [9]. A new spacetime framework was recently proposed to measure 4D displacements by combining DVC and P-DVC analyses [28]. By using spatiotemporal separations of variables, the spatial components were obtained via scan-wise DVC, and the *instantaneous* temporal amplitude was computed via P-DVC on each individual radiograph acquired on-the-fly during the whole loading history. In the following study, a new P-DVC enhanced DVC algorithm is introduced in which spatial modes are measured via DVC, and the temporal modes are sought with P-DVC. For the first time, the temporal modes are constructed such that they are compatible with the loading history of the experiment. In this regard, temporal shape functions are introduced. The displacement fields are sought in a 4D (i.e., 3D in space and 1D in time) vector space generated by a reduced spacetime kinematic basis. Only one projection per loading level is still needed. Additionally, different temporal interpolations are investigated, and the number of spatial modes selected to describe the specimen kinematics is analyzed and justified.

The aim of this work is the 4D characterization of a mat glass fiber reinforced polyester resin subjected to in situ cyclic tension, and imaged via XCT. A dogbone specimen contained a machined rectangular notch to induce high strain gradients. The properties and fabrication procedure of the investigated composite material are presented in Section 2, followed by a description of experimental setup. In Section 3, the principle of the proposed procedure is presented by introducing DVC and P-DVC formulations. The method is then applied to an in situ cyclic tensile test of a single-notched specimen (Section 4). The reconstruction error was first quantified by evaluating the difference between the corresponding projections and re-projections of the reconstructed reference volume. Last, the methodology to quantify damage growth from projection residual fields is presented. The implemented methodology is employed to directly assess damage from radiographs, thereby representing a completely novel approach.

2. Material and Methods

2.1. Material and Experimental Setup

In the present work, a polyester resin reinforced with a continuous glass fiber mat is investigated. The composite plate comprised 12 layers of R-glass fiber mat and was produced by manual lay-up, followed by compression molding. In terms of volume fraction, the composition had 40% of fibers, 55% of resin and 5% of voids (due to the fact that vacuum was not applied during molding). Figure 1 shows an optical micrograph of the investigated material. Due to the contrast between constituents, this material is well-suited for DVC, which relies on natural contrast [4].

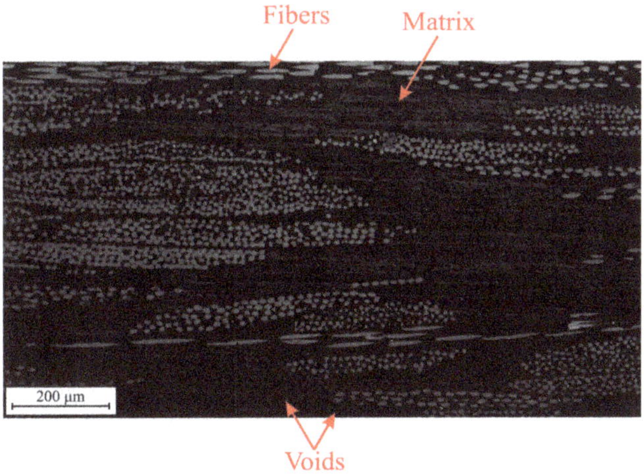

Figure 1. Optical micrograph of the glass fiber mat reinforced polyester resin composite studied herein.

The thickness of the investigated specimen was 5.2 mm. The specimen contained a machined rectangular notch (Figure 2a), which induced high strain gradients. The depth of the notch was 1.6 mm, while its width was 0.6 mm. The in situ cyclic tensile test was performed with the TTC Deben testing machine, i.e., the specimen was continuously loaded, rotated and imaged in the X50+ scanner (North Star Imaging) of LMPS. Before prescribing any loading, a reference scan (denoted with the black dots in Figure 2b), comprising 800 projections was acquired at equally spaced angles spanning over a complete 360° revolution with high-quality (HQ) scanning parameters (Table A1 in Appendix A). As this scan took approximately 2 h to be completed, the number of averaging frames was subsequently reduced from 20 to 1 to lower the experiment duration and mitigate, e.g., stress-relaxation. Such scanning parameters are called continuous scanning, and each scan took 4 min to be completed. After the acquisition of the reference scan, cyclic tension was applied at a constant stroke velocity of 4 μm/s. The radiographs were continuously acquired during the entire loading history at 4 frames per second with angular increments of 0.47°. These settings led to 768 radiographs per turn. It is worth highlighting that the P-DVC methodology presented herein is not dependent on the rate of the conducted experiments. However, there may be limitations related to the temporal resolution of each imaging device.

The deformed scans were reconstructed from the stages where a constant stroke was applied (marked with red dots in Figure 2b). They were used to measure 3D displacement fields via FE-DVC within the Correli 3.0 framework [29]. The FE mesh used in all DVC analyses (Figure 2a) was composed of first-order tetrahedral (T4) elements with piecewise linear (P1) shape functions. It was made of 88 nodes (with three DOFs per node), and the mean element length was equal to 12 vx.

A total of 17,442 projections was acquired during the prescribed loading history. The radiographs had an initial definition of 1944 × 1536 px. The focus was put on the notched-region (as the majority of damaged zones concentrated within this region [27]). Coarse graining of 4 × 4 elementary pixels into one superpixel was performed. This led to 486 × 384 px radiographs (called images at scale 4). The physical length of one voxel was equal to 58 µm. The volumes were reconstructed within the ASTRA toolbox with the simultaneous iterative reconstruction (SIRT) algorithm suited for cone beams [30], and employing geometrical parameters given by the tomograph calibration.

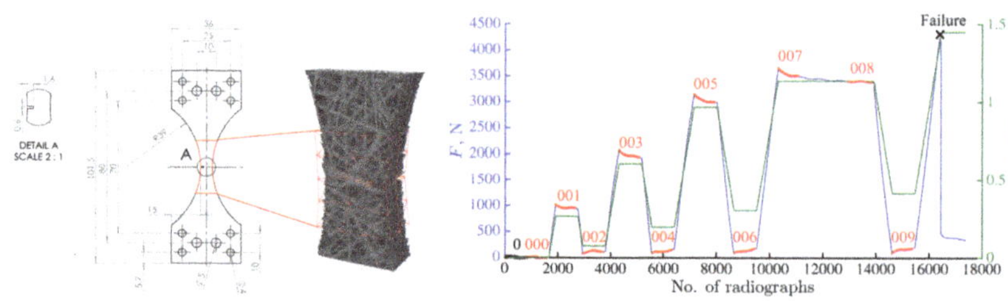

(a) (b)

Figure 2. (**a**) Geometry of the investigated single notched dogbone specimen together with the Region of Interest (depicted with red contour) and finite element mesh employed herein (superimposed over the reconstructed reference volume). The dimensions are expressed in mm. The size of Region of Interest was 384 × 384 × 486 vx, while the physical length of one voxel was 58 µm. (**b**) Measured uniaxial force (blue) and stroke history (green) of the studied in situ tensile test. The black dots mark the acquisition of the high-quality reference scan (0). The red dots depict the load levels at which full volumes were reconstructed. These volumes were employed in the scan-wise DVC analyses.

2.2. Projection-Based Digital Volume Correlation

This section outlines the basic principles of DVC and P-DVC. The notations used herein are introduced as well. The radiographs (i.e., the sum of absorption coefficient along each material point **x** of the beam ray hitting the detector at position **r**) are denoted as $p(\mathbf{r}, t)$ at different time t. The reconstructed volume is such that its projection should match the recorded projection $p(\mathbf{r}, t)$, which results in a linear relationship between f and p

$$\Pi_{\theta(t)}[f(\mathbf{x})] = p(\mathbf{r}, t), \tag{1}$$

where $\Pi_{\theta(t)}$ is the projection operator at angle $\theta(t)$. The reconstruction is thus the inversion of this linear system for a large sequence of angles [31]. By reconstructing a series of such tomographic images, the 3D kinematics of a medium can be quantified. The displacement field is estimated via DVC, which relies on the conservation of gray levels

$$f^{T_0}(\hat{\mathbf{x}}) = f^{T}(\hat{\mathbf{x}} + \hat{\mathbf{U}}(\hat{\mathbf{x}}, T)), \tag{2}$$

where $\hat{\mathbf{x}}$ denotes the position of any material point in the reference configuration, and $\hat{\mathbf{U}}$ the measured displacement field that describes the change of the reference volume f^{T_0} (usually captured in the undeformed state) to the deformed configuration (here denoted as f^T). Two different time scales have to be highlighted here. First, the 3D images are reconstructed over the duration needed for one turn T. Hundreds up to thousands of instantaneous radiographs acquired at $\theta(t)$ are needed over a full revolution (or turn T) of the specimen. The turn parameter T thus denotes the different stages at which a 3D scan was captured, while T^0 denotes that of the reference scan. The displacement field to be measured is obtained from the minimization of Γ_{DVC}, which is the quadratic differences between the corrected deformed and the reference volumes

$$\Gamma_{\text{DVC}} = \sum_{\hat{\mathbf{x}}} (f^T(\hat{\mathbf{x}} + \hat{\mathbf{U}}^{\text{DVC}}(\hat{\mathbf{x}}, T)) - f^{T_0}(\hat{\mathbf{x}}))^2 \tag{3}$$

with respect to the parameterization of the trial displacement fields $\hat{\mathbf{U}}^{\text{DVC}}$. A global (i.e., FE-based) approach was utilized in this work. The displacement field is then expressed in a kinematic basis consisting of FE shape functions $\hat{\mathbf{\Psi}}_j$

$$\hat{\mathbf{U}}^{\text{DVC}}(\hat{\mathbf{x}}, T) = \sum_j v_j(T) \hat{\mathbf{\Psi}}_j(\hat{\mathbf{x}}), \tag{4}$$

where v_j are the nodal displacements. As the sought displacement fields are determined with respect to the reference configuration, the DVC framework presented herein is Lagrangian.

Instead of working with fully reconstructed volumes, P-DVC measures 3D displacement fields from a series of 2D radiographs captured at different angles $\theta(t)$ and loading steps $F(t)$ [32]. The cost function $\Gamma_{\text{P-DVC}}$ to be minimized is defined as follows

$$\Gamma_{\text{P-DVC}} = \sum_{\mathbf{r},t} (\Pi_{\theta(t)}[f^{T_0}(\mathbf{x} - \mathbf{u}(\mathbf{x}, t))] - p(\mathbf{r}, t))^2. \tag{5}$$

According to Equation (5), the problem is no longer stated in the Lagrangian framework, but in the Eulerian system (contrary to Equation (4)). To complete the notations needed to describe the motion in Lagrangian and Eulerian settings, it is worth highlighting that the displacement at any spatial location $\mathbf{x}$ reads

$$\mathbf{u}(\mathbf{x}, t) = \mathbf{x} - \hat{\mathbf{x}} = \hat{\mathbf{u}}(\hat{\mathbf{x}}, t). \tag{6}$$

The present P-DVC framework requires the acquisition of one reference volume f^{T_0} from which the microstructure of the sample is known. The remaining unknowns are the displacements sought in the spatiotemporal framework. In the present study, the spatial components are constructed with all scan-wise DVC results, while only temporal modes are sought via P-DVC

$$\mathbf{u}(\mathbf{x}, t) = \sum_{\tau} \alpha_\tau \hat{\mathbf{U}}^{\text{DVC}}(\hat{\mathbf{x}}, \tau) \sigma_\tau(t), \tag{7}$$

where $\sigma_\tau(t)$ denote the temporal modes, and α_τ the temporal amplitudes. The presented methodology is 4D in the sense that it gives access to the entire displacement field $\mathbf{u}(\mathbf{x}, t)$ in (3D) space and (1D) time. Furthermore, DVC fields are the result of the minimization (Equation (3)) when the displacement fields are parameterized with FE shape functions (Equation (4)).

The displacement fields are sought in a vector space generated by a reduced kinematic basis. The interpolation functions σ_τ naturally introduce temporal regularization [9]. The nodal displacements $v_j(\tau)$ are measured via FE-DVC, and the temporal amplitudes gathered in column vector $\{\boldsymbol{\alpha}\}$ are sought via P-DVC and iteratively updated

$$\{\boldsymbol{\alpha}^{l+1}\} = \{\boldsymbol{\alpha}^l\} - \{\delta\boldsymbol{\alpha}^l\}, \tag{8}$$

where the corrections are calculated by performing Gauss-Newton minimizations of the functional $\Gamma_{\text{P-DVC}}$

$$\{\delta\boldsymbol{\alpha}^l\} = [\mathbf{M}^l]^{-1}\{\mathbf{m}^l\}. \tag{9}$$

In Equation (9), the Hessian matrix $M_{\tau k}$ is updated for each iteration l

$$M^l_{\tau k} = \sum_{\mathbf{r},t}(S^l_\tau(\mathbf{r},t)\sigma_\tau(t)) \cdot (S^l_k(\mathbf{r},t)\sigma_k(t)), \tag{10}$$

where S^l_τ is the projected sensitivity at iteration l

$$S^l_\tau(\mathbf{r},t) = \Pi_{\theta(t)}[\hat{\mathbf{U}}^{\text{DVC}}(\hat{\mathbf{x}},\tau) \cdot \nabla f^{T_0}(\mathbf{x} - \mathbf{u}^l(\mathbf{x},t))]. \tag{11}$$

Furthermore, the second member vector m^l_τ is based on the projection residual fields

$$m^l_\tau = \sum_{\mathbf{r},t} S^l_\tau(\mathbf{r},t)\sigma_\tau(t)\varphi^l_C(\mathbf{r},t), \tag{12}$$

where the projection residual field per angle at each iteration l reads

$$\varphi^l_C(\mathbf{r},t) = p(\mathbf{r},t) - \Pi_{\theta(t)}[f^{T_0}(\mathbf{x} - \mathbf{u}^l(\mathbf{x},t))]. \tag{13}$$

According to Equation (13), the reference volume, corrected by the displacement field $\mathbf{u}^l(\mathbf{x},t)$, and projected for each angle $\theta(t)$ should match the acquired projection $p(\mathbf{r},t)$. The projection residual field for all angles is to be minimized during the iterative procedure. As in DVC, these residuals reveal what was not captured by the 4D corrections, herein projected onto the 2D detector plane. A general overview of 4D analyses in which spatial modes are controlled by DVC fields is shown in Algorithm 1.

Algorithm 1: P-DVC with DVC spatial modes

Select spatial modes $\hat{\mathbf{U}}^{\text{DVC}}(\hat{\mathbf{x}},\tau)$
Choose an initial guess of $\{\boldsymbol{\alpha}^0\}$
while $\|\{\delta\boldsymbol{\alpha}^l\}\| > 10^{-3}$ **do**
  Corrections $f^l(\mathbf{x},t) \leftarrow f^{T_0}(\mathbf{x} - \mathbf{u}^l(\mathbf{x},t))$
  Update $[\mathbf{M}^l]$ and $\{m^l\}$
  Solve $\{\delta\boldsymbol{\alpha}^l\} = [\mathbf{M}^l]^{-1}\{\mathbf{m}^l\}$
  Update temporal amplitudes $\{\boldsymbol{\alpha}^{l+1}\} = \{\boldsymbol{\alpha}^l\} - \{\delta\boldsymbol{\alpha}^l\}$
  Update displacement fields $\mathbf{u}^{l+1}(\mathbf{x},t) = \sum_\tau \alpha^{l+1}_\tau \hat{\mathbf{U}}^{\text{DVC}}(\hat{\mathbf{x}},\tau)\sigma_\tau(t)$
  Update projection residuals $\varphi^{l+1}_C(\mathbf{r},t) = p(\mathbf{r},t) - \Pi_{\theta(t)}[f^{T_0}(\mathbf{x} - \mathbf{u}^{l+1}(\mathbf{x},t))]$
end

3. Results

In this section, the previous P-DVC enhanced DVC framework is applied to an in situ cyclic tensile test on a single-notched specimen. First, the reconstruction error of tomographic volumes is evaluated. The results of 4D measurements are then presented. All presented analyses converged, i.e., the norm of the change of amplitude corrections $\{\delta\boldsymbol{\alpha}\}$ (Equation (9)) between two iterations became less than 10^{-3}. Convergence and the obtained temporal amplitudes are discussed.

3.1. Evaluation of Reconstruction Error

According to Equation (13), the P-DVC algorithm minimizes the projection residual fields (i.e., differences between the acquired radiograph $p(\mathbf{r},t)$ and the reference volume deformed by the measured displacement field, and projected according to the angle $\theta(t)$).

This procedure is herein extended to the initial step (i.e., the reference HQ scan 0), where $\mathbf{u} = \mathbf{0}$. The reconstruction error was quantified by evaluating the difference between the corresponding projections and re-projections of the reconstructed reference volume 0. Due to acquisition noise, uncertainties in the geometric parameters used for the reconstruction and assumptions in the projection operator (e.g., pixel/voxel integration, beam hardening), the initial projection residual field was not equal to 0.

The root mean square (rms) of projection residuals (φ_C), expressed in arbitrary units, evaluated for the first 360° revolution (scan 0) is shown in Figure 3a. The projection residuals within the whole ROI are low with respect to the mean dynamic range of the original radiographs (i.e., 0.7 a.u.). The highest residual levels are reached for $\theta = 90°$ and $\theta = 270°$. Figure 3b,c shows the projection residuals for $\theta = 90°$ and $\theta = 270°$. These initial projection residual maps are shown with a divergent color map to highlight positive and negative values. The highest rms(φ_C) levels (i.e., $\theta = 90°$ and $\theta = 270°$) are due to the specimen orientation with respect to the X-ray source. The thickness of the specimen with respect to the X-ray source is the highest at those angles, which may introduce nonlinearity in the X-ray attenuation. The initial projection residual field for $\theta = 0°$ reveals elevated levels close to the edges of the notch. The present volumes were reconstructed with the SIRT algorithm, which iterates between forward and backprojections until it reaches convergence. The algorithm was initialized with the Feldkamp–Davis–Kress (FDK) procedure. SIRT is computationally more expensive compared to FDK. However, it provides images with reduced noise. In the present case, the volumes were reconstructed with 100 SIRT iterations. The reconstruction residuals may be reduced by further adjusting the geometric parameters of the tomographic setup, and increasing the number of SIRT iterations. Other sources for such errors could be small motions during scanning and beam hardening, which were not corrected in the present work.

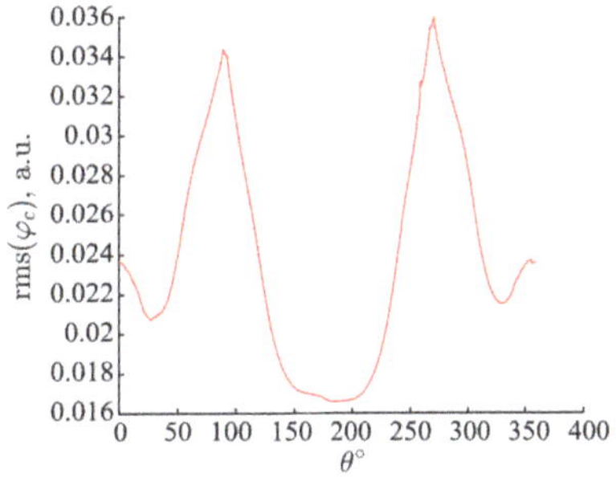

(a)

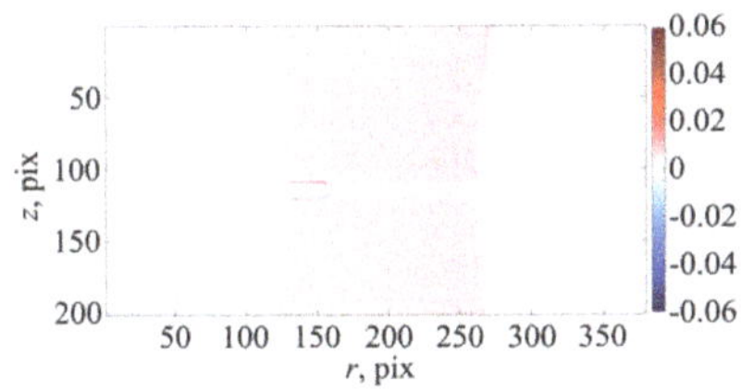

(b)

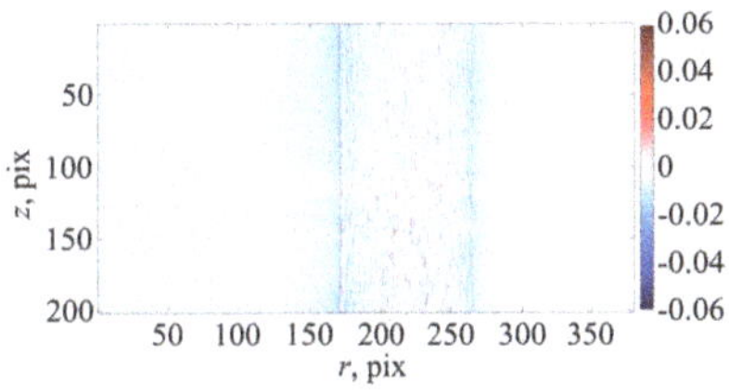

(c)

Figure 3. (**a**) Root mean square (rms) projection residuals evaluated for the first 360° revolution (scan 0). Initial projection residual field for $\theta = 0°$ (**b**) and $\theta = 90°$ (**c**).

3.2. Full-Field Measurements over the Entire Loading History

All the results presented in the sequel are obtained with a temporal basis consisting of 11 plateau-like functions interpolating the cyclic loading history (Figure 4). The selection of this temporal basis is detailed in Appendix B. The temporal shape functions are linked to 11 spatial modes. The latter ones are constructed via DVC, i.e., by performing volumetric correlations between scans 0 (reference scan) and all subsequent reconstructed volumes (Figure 2b). It is important to emphasize that all P-DVC measurements are performed until specimen failure (i.e., *after* acquiring scan 009). The last temporal mode is related to the DVC field of scan 007. The (small) force drop between scans 007 and 008 was due to the fact that the continuous acquisition between these two steps was interrupted by acquiring one high quality scan, which was excluded from the present analyses.

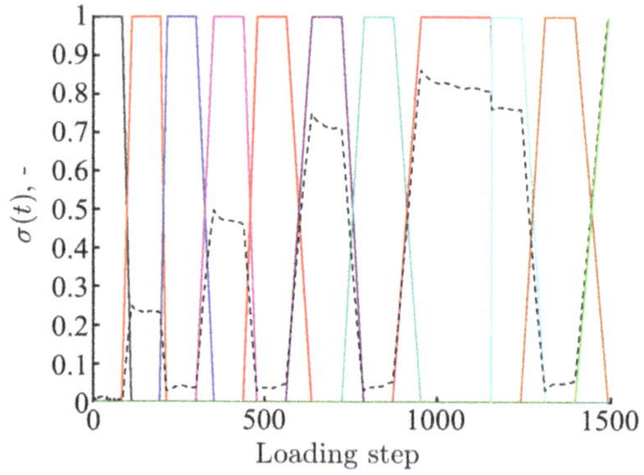

Figure 4. Temporal modes consisting of 11 plateau-like functions. The black dashed line represents the measured force signal normalized by its maximum value.

The projections to be registered with P-DVC are selected so that the angle between two consecutive acquisitions is approximately equal to 4.6°. This choice results in 1491 projections to be analyzed (out of a total of 17,442 acquired during the complete loading history). The chosen temporal sampling leads to 77 projections per turn (360° revolution). The sampling is governed by the computation time. In P-DVC procedures, the heaviest operations are the computation of the Hessian matrices and the deformation of the reference volume at each time step. To perform measurements in reasonable time, such trade-off had to be made between temporal resolution and computation time. Compared to standard DVC analyses that would yield 11 fields in the present analyses, the full P-DVC analysis probes 1491 radiographs corresponding to different temporal states (i.e., more than 2 orders of magnitude higher than regular DVC).

The initial values in the amplitude vector $\{\boldsymbol{\alpha}^0\}$ are set to 1 (Equation (7)). The P-DVC analysis takes 4 iterations to converge. Each iteration lasts approximately 5 h. The measured displacements yield a global decrease of the residuals within the investigated ROI as shown in Figure 5.

The converged temporal amplitudes are gathered in Table 1. They are all close to 1, except for the last one where the amplitude is greater than 1. This is due to the fact that the DVC displacement field of scan 007 (Figure 2b) is used as a spatial component of the last mode, while 4D kinematic measurements are performed up to specimen failure.

Furthermore, the rms differences between the 3D nodal displacements obtained with FE-based DVC and projection-based measurements are calculated. Since large RBRs in x and y directions are observed during the experiment, these differences are only calculated for z-displacements as the mechanical component is dominant in that direction. The rms differences are computed between the DVC displacement fields and converged P-DVC displacement fields for the time-steps corresponding to the beginning of each constant stroke stage. The corresponding values are gathered in Table 2. As the projections acquired at the maximum load levels are affected by time-dependent motions (i.e., stress-relaxation phenomena), which influence the reconstruction of the full volume, differences are expected. From the values reported in Table 2, it is concluded that, when employing 11 plateau-like temporal functions, the rms differences are very small. With such settings, the kinematics of the specimen was properly captured.

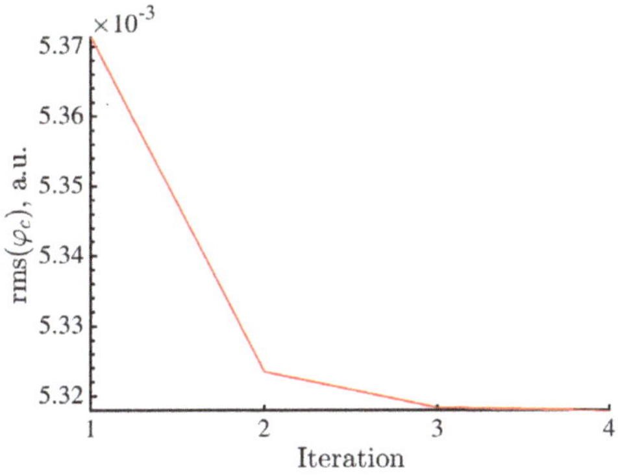

Figure 5. Change of rms projection residuals over the entire Region of Interest during the iterative procedure.

Table 1. Converged temporal amplitudes obtained when using 11 plateau-like functions as the temporal basis. The * highlights an extrapolation, namely, the temporal interpolation function describing the last loading stage prior to failure is related to the spatial mode of scan 007.

Scan	Mechanical State	α
000	Unloaded	1.005
001	Loaded	1.007
002	Unloaded	0.983
003	Loaded	1.007
004	Unloaded	1.012
005	Loaded	1.002
006	Unloaded	0.998
007	Loaded	1.011
008	Loaded	1.006
009	Unloaded	0.996
007 *	Loaded	1.128

Table 2. Comparison of rms differences expressed in vx in z-direction between the nodal displacements measured with FE-DVC and projection-based measurements. The comparison is performed for the converged P-DVC displacement fields for the angles at the beginning of each constant stroke stage.

Scan	Mechanical State	z
000	Unloaded	0.005
001	Loaded	0.007
002	Unloaded	0.01
003	Loaded	0.007
004	Unloaded	0.012
005	Loaded	0.002
006	Unloaded	0.011
007	Loaded	0.012
008	Loaded	0.011
009	Unloaded	0.011

4. Discussion: Quantification of Damage Growth

In this section, the methodology to quantify damage growth from projection residual fields is outlined and discussed for one particular case. The validation of the proposed procedure is detailed in Appendix C. The starting point of the method is to exploit the fact that DVC residuals revealed damaged zones within the specimen. Figure 6a shows the DVC residual field of scan 007 (last loading plateau prior to specimen failure). This map reveals numerous damaged zones in the specimen bulk, the majority of which are concentrated within the notched region. Next, the absolute maximum value of correlation residuals within the whole ROI is determined. The following criterion is employed to find finite elements that have correlation residuals greater than 70% of the maximum value integrated over each element volume

$$\mathrm{rms}(\varphi_{C,\mathrm{DVC}})^{\mathrm{Mask}} > 0.7 \cdot \mathrm{rms}(\varphi_{C,\mathrm{DVC}})^{\mathrm{ROI,max}} \tag{14}$$

Based on the previous criterion, 14 elements are found. These elements compose the mask shown in Figure 6b, which is used to inspect the projection residuals. The mask is projected onto the detector plane with the projection operator $\Pi_{\theta(t)}$. The quantification of damage growth is performed within each element of the mask.

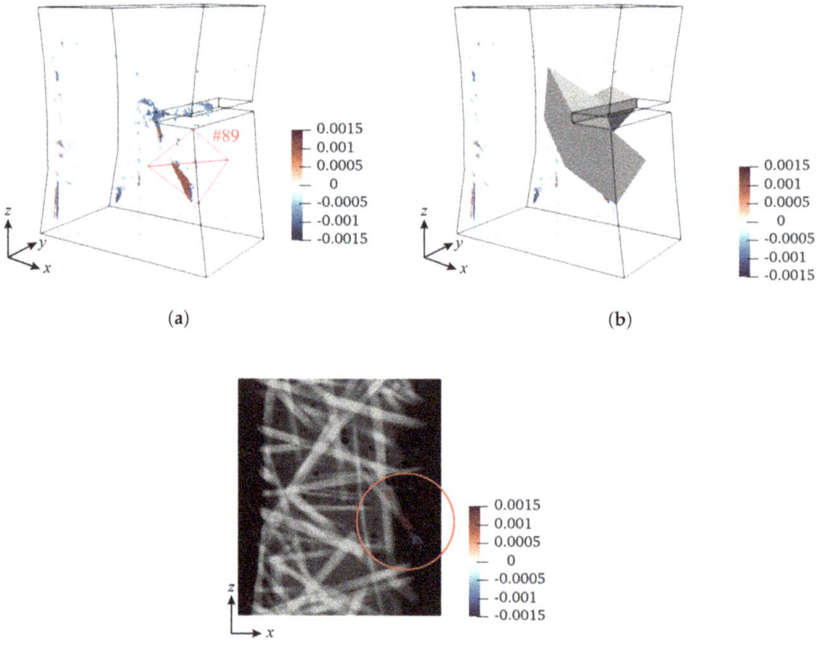

(c)

Figure 6. (**a**) DVC residuals of scan 007. The red contour denotes element #89. (**b**) Finite elements containing the highest DVC residual levels. (**c**) DVC gray level residuals laid over the mesostructure section of the front specimen surface. The red circle marks the area exhibiting elevated correlation residuals within element #89.

The analysis of damage growth is performed within element #89, which is located below the notch root (marked with red contour in Figure 6a. This element contains the zone of elevated correlation residuals, which correspond to a crack propagating from the notch root (Figure 6c).

As 1491 radiographs are analyzed with P-DVC, not all angles are inspected in the damage procedure. To find for which angles damage (i.e., mesocrack) is most pronounced, the DVC residuals within element #89 are projected onto the detector plane. First, the rms residuals are computed for the full specimen revolution for both loaded and unloaded plateaus. The residuals of the loaded scans are shown in Figure 7a, whereas those of the unloaded acquisitions in Figure 7b. The rms residual levels of scan 000 (acquired in the undeformed state) is included in order to track the increases in DVC residuals with respect to the base level. For scan 001, there is an increase in correlation residuals with respect to levels of scan 000. When the specimen is unloaded (scan 002), the residuals remain close to the levels of scan 001. This is trend is also visible when plotting the rms DVC residuals of element #89 during the entire returns for the acquisitions of scans 001 and 002. These increases in correlation residuals compared to the base level are attributed to the specimen kinematics [27] and damage still did not occur at this stage.

As the specimen is once again loaded (scan 003), the correlation residuals keep increasing. Upon subsequent unloading, the residual levels decrease and are very close to those of scan 002. Damage growth did not yet occur, or if it had, it remained very low and at a scale lower than the scan resolution [33]. The significant increase in the correlation residuals compared to previous acquisitions is visible for scan 005. This is especially pronounced in Figure 7c. At this stage, the crack originating from the notch root and propagating below it is visible on DVC correlation residual field [34]. The subsequent unloading (scan 006) led to partial crack closure (as the correlation residuals were higher than those of the previous unloaded acquisitions, Figure 7c). At the peak of the last loading plateau prior to specimen failure (scans 007 and 008), the correlation residuals once again increase, and upon specimen unloading (scan 009), partial crack closure appears. From the reported data, the first interval of residual increase is within the range of 50° to 110°. In this range, damage is expected to be pronounced on residual fields. Thus, the projected DVC residuals within element #89 are shown in Figure 7d for an angle of 75°. The area containing elevated residuals corresponding to the crack is marked with the red arrow.

With the previous observations, the projection residuals are only inspected for specific angles, namely, 0 (to prove that this angle is damage-insensitive for this specific case), 56°, 75° and 105° acquired during the loading history. First, damage growth is analyzed by tracking the increase in rms($\varphi_C^{\text{P-DVC}}$) with respect to the base level. The rms projection residual levels are divided by that of scan 000 (prior to loading). This is the so-called amplification factor shown in Figure 8.

For $\theta = 0°$, there is no increase that may reveal damage growth, except for the very last loading step prior to specimen failure, which may also be attributed to the extrapolation using the DVC displacement field of scan 007. The same conclusions are drawn for $\theta = 56°$. Furthermore, the amplification levels when $\theta = 75°$ reveal the first significant increase in projection residuals compared to the base level for time step 941, which corresponds to the range of scan 007 (Figure 2b). This observation is consistent with DVC results at a finer scale [34].

The accumulation of damage and the activation and deactivation of cracks (due to the cyclic loading history) is most visible for $\theta = 105°$. It is concluded that damage within element #89 first initiated at time step 331, which corresponds to the range of scan 003 (Figure 2b). During the subsequent unloading (time step 485), it appears that damage is deactivated. Damage within element #89 is once again active upon loading the specimen to the peak of the third loading cycle (time step 639). However, when the specimen is once again unloaded (time step 793), the crack within element #89 is only partially closed (the amplification of time step 793 is greater than that of time step 485). Within the range of scans 007 and 008 (time steps 947 to 1255) the crack is again active and more pronounced compared to the previous loading stages. The subsequent unloading (time step 1332), once again leads to partial crack closure.

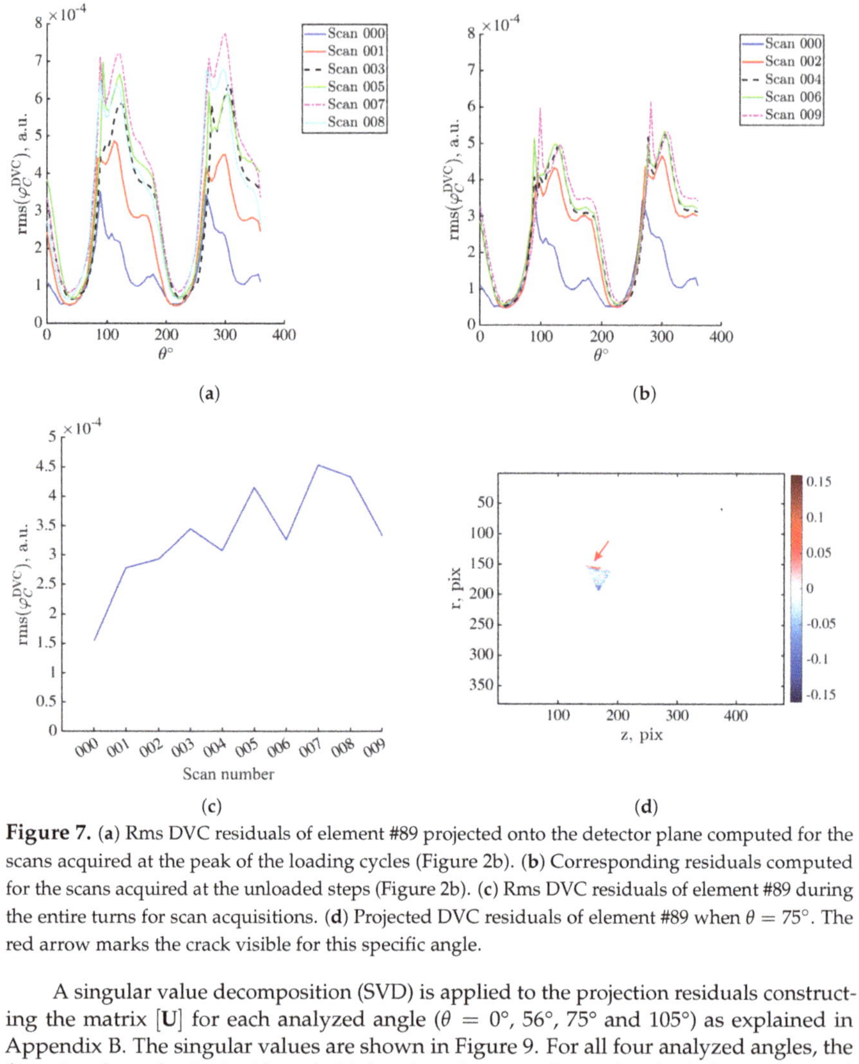

Figure 7. (**a**) Rms DVC residuals of element #89 projected onto the detector plane computed for the scans acquired at the peak of the loading cycles (Figure 2b). (**b**) Corresponding residuals computed for the scans acquired at the unloaded steps (Figure 2b). (**c**) Rms DVC residuals of element #89 during the entire turns for scan acquisitions. (**d**) Projected DVC residuals of element #89 when $\theta = 75°$. The red arrow marks the crack visible for this specific angle.

A singular value decomposition (SVD) is applied to the projection residuals constructing the matrix $[\mathbf{U}]$ for each analyzed angle ($\theta = 0°$, 56°, 75° and 105°) as explained in Appendix B. The singular values are shown in Figure 9. For all four analyzed angles, the first singular value is significantly higher than the second one, thereby indicating a good separation between them. Thus, only the first temporal mode of each angle is analyzed for the purpose of quantification of damage growth.

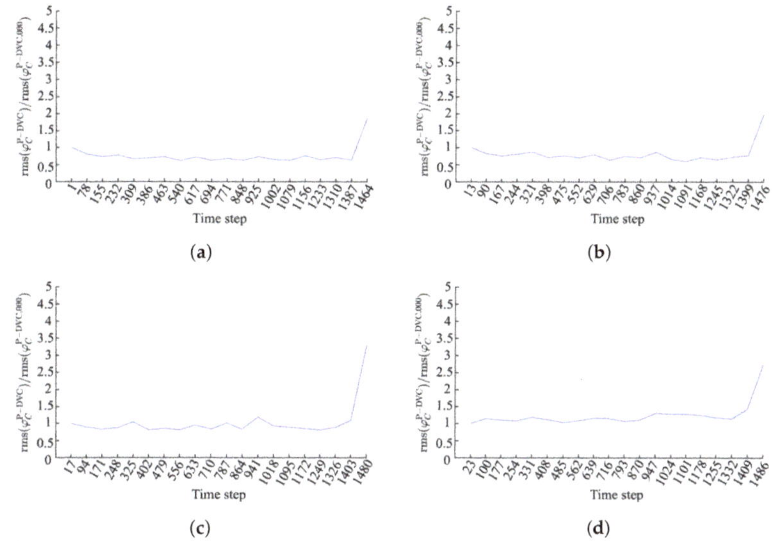

Figure 8. Amplification of projection residuals within element #89 when (**a**) $\theta = 0°$, (**b**) $\theta = 56°$, (**c**) $\theta = 75°$ and (**d**) $\theta = 105°$.

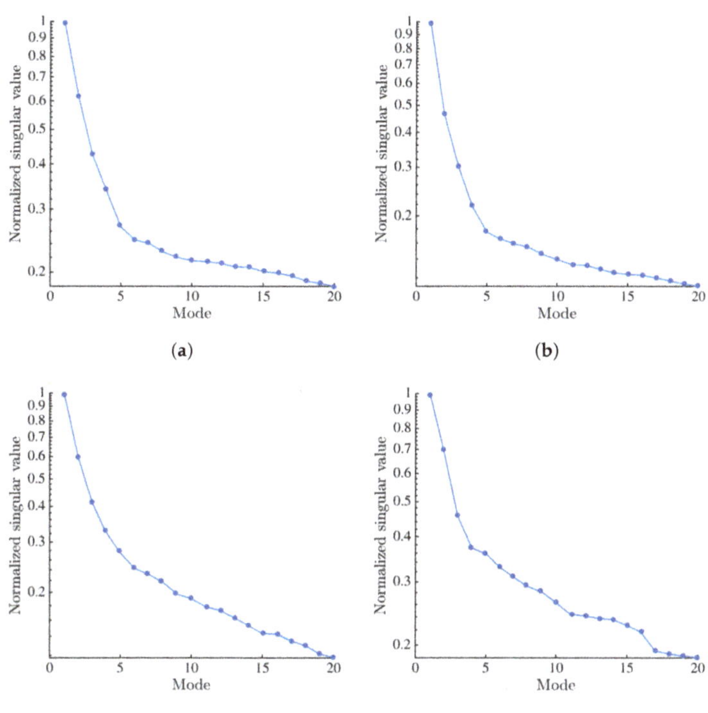

Figure 9. Normalized singular values for element #89 when (**a**) $\theta = 0°$, (**b**) $\theta = 56°$, (**c**) $\theta = 75°$ and (**d**) $\theta = 105°$.

Figure 10 shows the first temporal modes calculated when $\theta = 0°$, 56°, 75° and 105°. For $\theta = 0°$ and 56°, the values for different loading steps (in this case marked with turns) are distributed within the same range. Thus, for these angles, damage is not detected. The increases in values for the last turn prior to specimen failure mainly stem from the extrapolation of the DVC field of scan 007. Furthermore, the values obtained for $\theta = 75°$ reveal the first increase for the loading stage corresponding to scan 005 (turns 9 and 10). However, for subsequent unloading the crack is only partially closed. This remark is consistent with DVC results on a finer scale [34]. Damage growth is pronounced for scans 007 and 008 (turns 13 and 14). The subsequent unloading resulted in partial crack closure. The values reported for $\theta = 105°$ clearly highlight the accumulation of damage within element #89. These values also reveal the activation and deactivation of damage. Thus, this angle is deemed the most sensitive to the presence of cracks.

The first spatial modes of the projection residuals computed for the selected angles are shown in Figure 11. For $\theta = 0°$, 56° and 105°, the spatial distribution is homogeneous. The area of elevated values is visible near the element angle when $\theta = 75°$. The reported results demonstrate that the analysis of projection residuals enables for the quantification of damage growth within the investigated specimen. However, due to the heterogeneous and random mesostructure, the damage analysis is highly dependent on the analyzed angles.

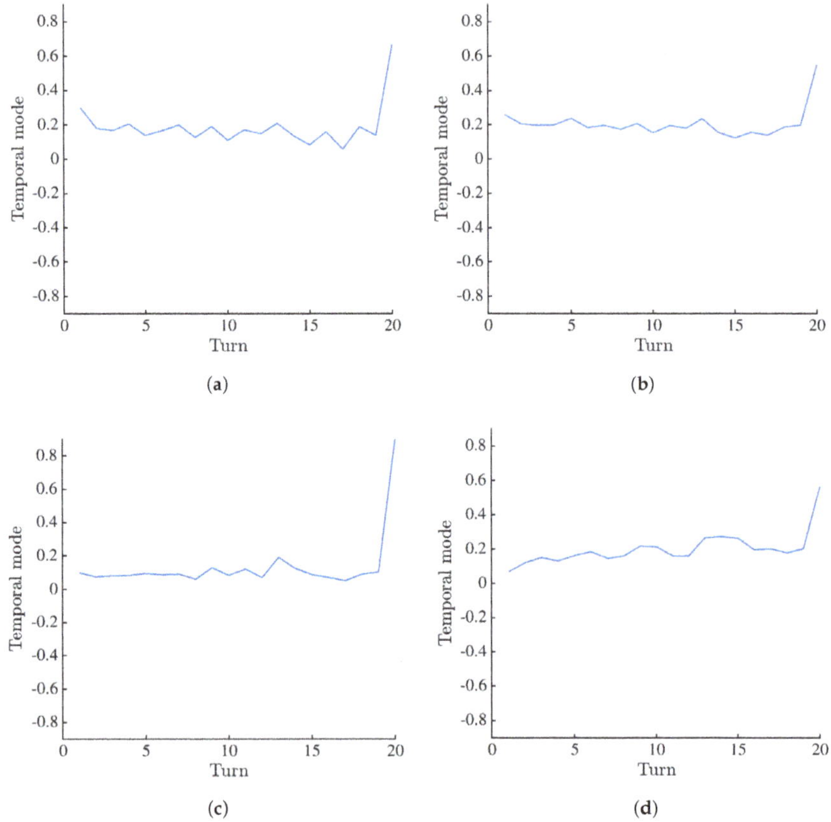

Figure 10. First temporal mode of projection residuals within element #89 computed when (**a**) $\theta = 0°$, (**b**) $\theta = 56°$, (**c**) $\theta = 75°$ and (**d**) $\theta = 105°$.

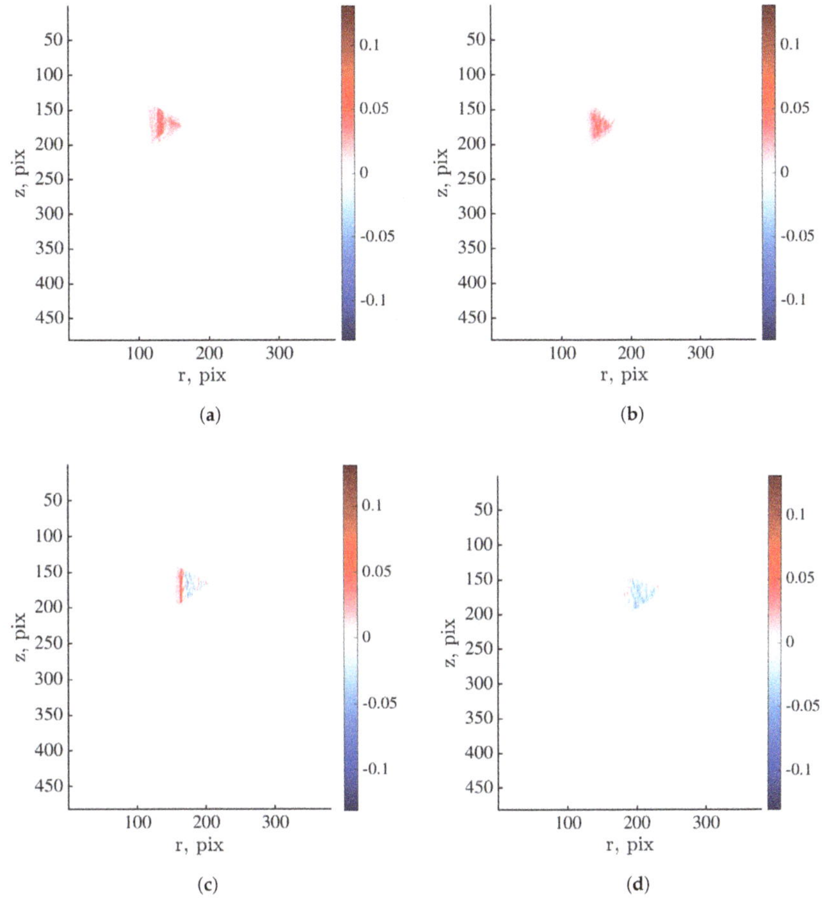

Figure 11. First spatial mode of projection residuals within element #89 calculated when (**a**) $\theta = 0°$, (**b**) $\theta = 56°$, (**c**) $\theta = 75°$ and (**d**) $\theta = 105°$.

5. Conclusions

In the present work, a 4D characterization of a glass fiber mat reinforced polyester resin was performed by employing DVC enhanced P-DVC that relied on spacetime discretization of the measured displacement fields. An alternative approach in which spatial modes were fully constructed with DVC, and only temporal modes were sought via P-DVC was implemented. All the analyses were performed with a temporal basis consisting of 11 plateau-like functions interpolating the cyclic loading history. These temporal shape functions were linked to 11 spatial modes constructed via DVC (i.e., by performing volumetric correlations between scan 0 and all subsequent reconstructed volumes).

A methodology to quantify damage growth was proposed by analyzing projection residuals for specific angles. First, zones in which damage growth was to be quantified were selected based on the DVC gray level residuals. Finite elements exhibiting correlation residuals greater than 70% of the maximum value within the entire ROI composed the mask. This mask was then projected onto the detector plane. In the present study, the analysis of damage growth was performed for one element that contained elevated correlation residuals corresponding to crack propagating from the notch root. The quantification of damage growth was performed by multiplying the projection residuals of this element with the mask of this element. As the temporal sampling was such that 1491 radiographs were analyzed with P-DVC, not all angles were inspected. To find for which angles damage was most pronounced, the DVC gray level residuals of the same element were projected onto the detector plane. Based upon the root mean square residual, specific angles were analyzed. Damage growth was first analyzed by tracking the increase in terms of rms values with respect to the base level. An amplification factor that revealed the increase in residual levels during the cyclic loading history indicated the activation and deactivation of damage. Last, an SVD procedure was also applied to the projection residuals. It was shown that the analysis of projection residuals may enable for the quantification of damage growth within the investigated specimen.

One of the principal aspects of future investigations will be to modify the temporal sampling in a way to find angles most sensitive to cracks. In such a way, the angles that do not display damage would be excluded from the analysis. Further, the presented results were obtained using coarse grained images (i.e., the resulting definition was down-sampled over a 4×4 regular grid). This binning level may lead to a loss of information beneficial for mechanical characterization. Another aspect will thus be to perform 4D measurements at the original scale. In addition, the measurements may be improved by further adjusting the geometrical parameters of the tomography setup, as well as performing beam hardening corrections (which was not carried out in the current work).

The P-DVC procedure employed in the present work offers the opportunity to continuously measure 4D (i.e., 3D space and 1D time) displacement fields with simultaneous loading and rotation of specimens. In the proposed framework, the spatial modes were constructed with scan-wise DVC analyses, and only temporal modes were sought via P-DVC. The P-DVC enhanced DVC method enabled for the quantification of damage growth over the entire loading history up to failure. The continuous acquisition of radiographs over the entire loading history and the analysis of only one radiograph per loading step significantly enriched the temporal sampling compared to scan-wise DVC (i.e., by more than two orders of magnitude in the present case). Due to its rich temporal sampling, the proposed framework may also enable for the assessment of time-dependent phenomena such as, e.g., stress-relaxation and crack propagation. It is important to highlight that the employed P-DVC methodology is generic in a way that it can be applied to different materials with various damage mechanisms and behavior, as well as different in situ experimental protocols (when specimens are simultaneously loaded and imaged). In addition to enabling experimental tests to be performed very quickly, it also allows different in situ experimental protocols (designed for standard full-field measurements) to be optimized. This procedure may significantly reduce the quantity of data required for volumetric full-field measurements, while enriching the temporal sampling compared to standard DVC approaches.

Author Contributions: Conceptualization, A.M., F.H. and Z.T.; methodology, C.J., B.S. and F.H.; software, A.M., C.J. and V.K.; validation, V.K., C.J. and F.H.; formal analysis, A.M., V.K. and F.H.; data curation, B.S.; writing—original draft preparation, A.M.; writing—review and editing, V.K., C.J., Z.T. and F.H.; supervision, F.H. and Z.T.; project administration, Z.T.; funding acquisition, Z.T. All authors have read and agreed to the published version of the manuscript.

Funding: This work was performed within the FULLINSPECT project supported by the Croatian Science Foundation (UIP-2019-04-5460 Grant). This work was also partially supported by the French "Agence Nationale de la Recherche" through the "Investissements d'avenir" program (ANR10-EQPX-37 MATMECA Grant). The financial support of the French-German University through the French-German Doctoral college "Sophisticated Numerical and Testing Approaches" (CDFA-DFDK 19-04) is acknowledged.

Data Availability Statement: The dataset is currently unavailable.

Conflicts of Interest: The authors declare no conflict of interest. The funders had no role in the design of the study; in the collection, analyses, or interpretation of data; in the writing of the manuscript; or in the decision to publish the results.

Abbreviations

The following abbreviations are used in this manuscript:

DVC	Digital Volume Correlation
FE	Finite Element
FDK	Feldkamp–Davis–Kress
FRP	Fiber Reinforced Polymer
HQ	High Quality
P-DVC	Projection-based Digital Volume Correlation
RBM	Rigid Body Motion
RBR	Rigid Body Rotation
rms	Root mean square
ROI	Region Of Interest
SIRT	Simultaneous Iterative Reconstruction
SVD	Singular Value Decomposition
TTC	Tension, Torsion and Compression
XCT	X-ray Computed Tomography
4D	Four-dimensional (3D space and 1D time)

Appendix A. Hardware Parameters

Table A1. Scanning parameters.

Tomograph	North Star Imaging X50+	
X-ray source	XRayWorX XWT-240-CT	
Target/Anode	W (reflection mode)	
Filter	None	
Voltage	150 kV	
Current	7.8 µA	
Focal spot size	5 µm	
Tube to detector	910 mm	
Tube to object	53 mm	
Detector	Dexela 2923	
Definition	1944 × 1536 pixels (2 × 2 binning)	
Scanning settings	High quality	Continuous
Number of projections	800	768
Angular amplitude	360°	360°
Frame average	20 per projection	Continuous (1 per step)
Frame rate	3 fps	3 fps
Acquisition duration	1 h 46 min 26 s	4 min 26 s
Reconstruction algorithm	SIRT	SIRT
Field of view	7.3 × 10 × 20.5 mm^3	7.3 × 10 × 20.5 mm^3
Image scale	14.6 µm/voxel	14.6 µm/voxel

Appendix B. Selection of Temporal Bases

In this appendix, the selection of a suited temporal basis is discussed. This analysis is first conducted by performing a singular value decomposition (SVD) on the DVC displacement fields. An SVD is applied to rectangular matrices $[\mathbf{U}]_{m\times n}$, with $m > n$ in this work. To construct this matrix, the nodal displacements of each DVC result are gathered in a column vector, and then concatenated to construct a rectangular matrix $[\mathbf{U}]$ that corresponds to a space/turn separation of the measured data.

The mesh consists of 88 nodes (with 3 DOFs per node), and 10 DVC fields are used (namely, scans 000 to 009). Therefore, a 264 × 10 $[\mathbf{U}]$ matrix is created. Three matrices are generated after conducting SVD: the singular values matrix $[\mathbf{S}]_{m\times n}$, and the modal matrices $[\mathbf{V}]_{m\times m}$ and $[\mathbf{T}]_{n\times n}$. These matrices allow for the reconstruction of matrix $[\mathbf{U}]$

$$[\mathbf{U}] = [\mathbf{V}][\mathbf{S}][\mathbf{T}]^\top, \tag{A1}$$

where $[\mathbf{S}]$ is a rectangular matrix, and the non-zero terms of this matrix are the singular values s_i of the matrix ordered by importance ($s_1 \geq s_2 \geq s_3 \geq \ldots s_n$). Due to their ranking, this matrix indicates that certain modes are more significant than others by several orders of magnitude. The spatial modes (SM) are gathered in matrix $[\mathbf{V}]$, while the temporal modes (TM) are gathered in $[\mathbf{T}]$. These modes can be truncated to obtain approximations $[\mathbf{U}_k]$ with the first k instead of all n modes

$$[\mathbf{U}_k] = \sum_{j=1}^{k} s_j \{\mathbf{V}_j\}\{\mathbf{T}_j\}^\top, \tag{A2}$$

where s_j is the j-th singular value $\{\mathbf{V}_j\}$ and $\{\mathbf{T}_j\}$ is the corresponding modal vectors.

Furthermore, the displacement fields measured with DVC consisted of rigid body motions (RBMs) $\mathbf{U}^{\mathrm{DVC}}_{\mathrm{rbm}}(\hat{\mathbf{x}}, T)$ and the part causing the deformation called mechanical displacement $\mathbf{U}^{\mathrm{DVC}}_{\mathrm{mec}}(\hat{\mathbf{x}}, T)$

$$\hat{\mathbf{U}}^{\mathrm{DVC}}_{\mathrm{tot}}(\hat{\mathbf{x}}, T) = \hat{\mathbf{U}}^{\mathrm{DVC}}_{\mathrm{rbm}}(\hat{\mathbf{x}}, T) + \hat{\mathbf{U}}^{\mathrm{DVC}}_{\mathrm{mec}}(\hat{\mathbf{x}}, T). \tag{A3}$$

The SVD procedure is applied to both $\hat{\mathbf{U}}^{\mathrm{DVC}}_{\mathrm{tot}}$ (i.e., the displacement fields containing RBMs) and $\hat{\mathbf{U}}^{\mathrm{DVC}}_{\mathrm{mec}}$ (i.e., the displacement fields from which RBMs are subtracted). This analysis is performed by using the displacement fields of scans 000 to 009. Figure A1a shows the spectrum of singular values normalized by the highest one (i.e., s_1). From the reported data, it is shown that for both analyzed cases s_1 is greater than s_2 by about one order of magnitude, which indicates that Mode 1 captures most of the 4D kinematics. In the case of $\hat{\mathbf{U}}^{\mathrm{DVC}}_{\mathrm{tot}}$, the difference between the first two singular values was smaller compared to $\hat{\mathbf{U}}^{\mathrm{DVC}}_{\mathrm{mec}}$. This may be attributed to the influence of RBMs. It is worth highlighting that large RBMs are observed during the experiment, and hence they participate actively to the first mode of $\hat{\mathbf{U}}^{\mathrm{DVC}}_{\mathrm{tot}}$.

The temporal modes obtained with $\hat{\mathbf{U}}^{\mathrm{DVC}}_{\mathrm{tot}}$ are shown in Figure A1b. From $\mathbf{T}^1_{\mathrm{tot}}$, a clear temporal signature resembling the cyclic loading history is observed (Figure 2b). Because of the orthogonality requirement of SVD, the fluctuations for higher order modes are more difficult to interpret.

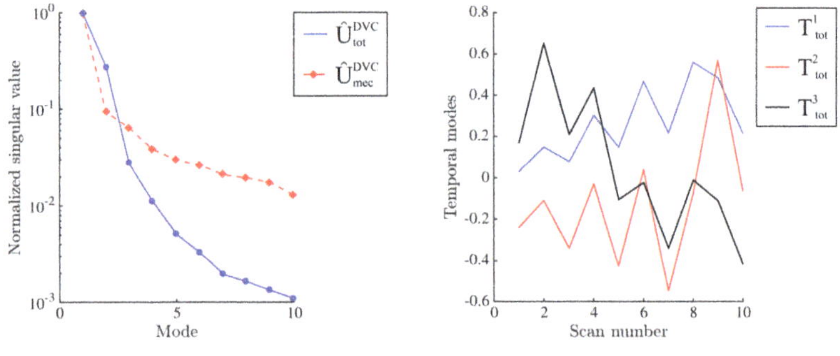

Figure A1. (**a**) Spectrum of normalized singular values. (**b**) First three temporal modes.

The first three spatial modes of $\hat{\mathbf{U}}^{\mathrm{DVC}}_{\mathrm{tot}}$ are shown in Figure A2. In the first two spatial modes, large rigid body rotations (RBRs) are observed in the in-plane components, which occurred during the experiment. For the loading direction, the displacements of the first two modes correspond to tensile loading of the notched sample. The third mode is much more difficult to elucidate.

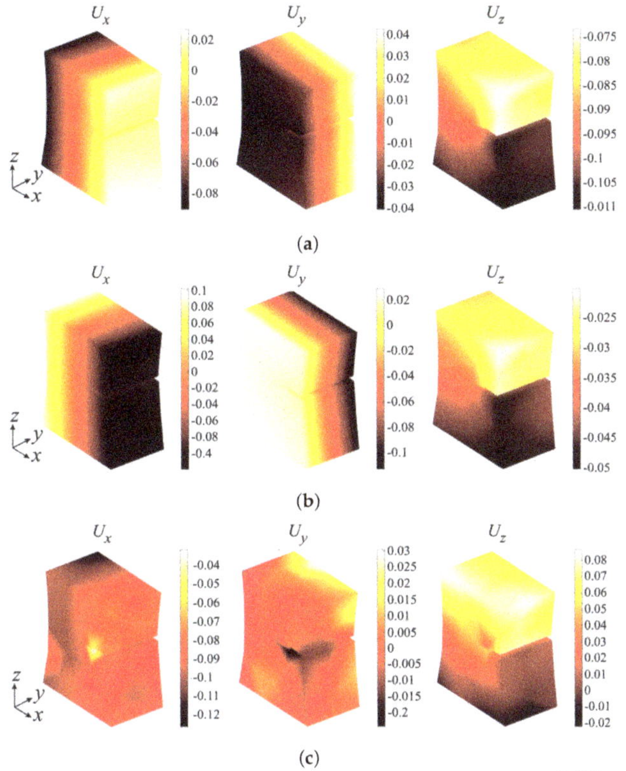

Figure A2. First (**a**), second (**b**), and third (**c**) spatial modes of $\hat{\mathbf{U}}^{\mathrm{DVC}}_{\mathrm{tot}}$.

The first two modes are compared to the displacement field of scan 007 (Figure 2b), as shown in Figure A3. The first mode is easily interpreted as the dominant kinematic mode.

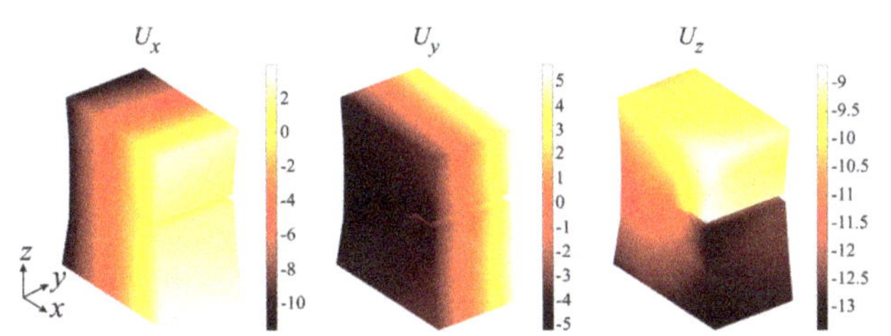

Figure A3. Displacement field of scan 007 containing RBMs.

To confirm the previous statement, Table A2 gathers rms differences between the DVC displacement field of scan 007 and rescaled 1st, 2nd, and 3rd spatial modes. The difference for the first spatial modes is very small compared to those of the two higher modes.

Table A2. Rms difference expressed in vx between the DVC displacement field of scan 007 and rescaled 1st, 2nd, and 3rd spatial mode.

	x	y	z
1st mode	0.34	0.08	0.09
2nd mode	4.6	2.0	1.2
3rd mode	4.4	2.7	7.8

Last, the truncation error evaluates the loss of information caused by using a subgroup of modes to reconstruct the kinematics. It is computed as

$$\text{Truncation error} = \frac{||[\mathbf{U}] - [\mathbf{U}_k]||^2}{||[\mathbf{U}]||^2}. \tag{A4}$$

Figure A4 shows the truncation error as a function of mode number. If only one mode is kept, the truncation error is significantly larger compared to the error if more modes are kept. For the first mode, when the total displacement field is considered, the truncation error is significantly higher than that of the mechanical field. This difference is due to significant RBRs. Since the truncation error when only 1 mode is kept is not too high for the total displacement fields, it is decided to investigate two reduced bases consisting of one single mode. Both of them are compared to results obtained with the full set of DVC displacements.

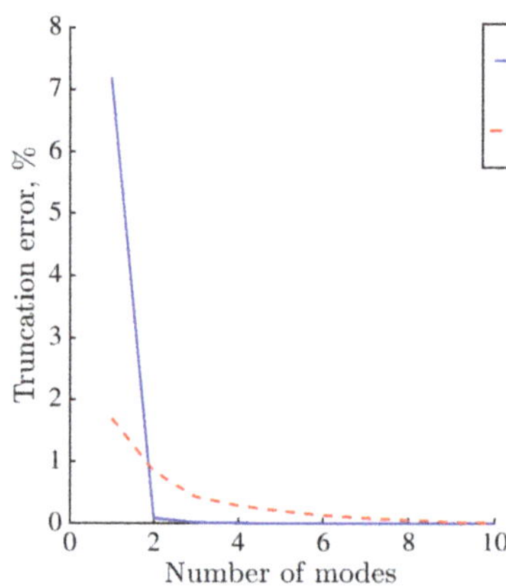

Figure A4. Truncation error as a function of mode number.

Different temporal interpolations are exploited on the chosen temporal sampling. First, the 4D kinematics is measured using the normalized force signal as a unique temporal mode (Figure A5a). The measured uniaxial force history reveals rapid but limited relaxations during the acquisition of scans 001, 003, and 005 (the drop of the peak force is approximately 160 N, see Figure 2b). During the acquisition of scans 007 and 008, the force decrease is approximately 50 N and 20 N, respectively.

Next, the normalized stroke is used as a unique temporal mode (Figure A5a). In both cases, only one DVC field is used as spatial mode, namely, the total displacement field of scan 007 (corresponding to the last loading cycle prior to specimen failure). This field is obtained by running DVC between HQ scan 0 (reference) and scan 007. It is important to emphasize that 4D measurements are performed until specimen failure (i.e., after acquiring scan 007). This extrapolation is justified by the fact that scan 007 is acquired at the peak of the last loading cycle prior to specimen failure.

The third approach consists in using 11 plateau-like functions interpolating the cyclic loading history (Figure A5). These temporal shape functions are linked to 11 spatial modes. The latter ones are constructed via DVC, i.e., by performing volumetric correlations between scans 0 (reference scan) and all subsequent reconstructed volumes (Figure 2b). It is important to emphasize that all P-DVC measurements are performed until specimen failure (i.e., after acquiring scan 009). The last temporal mode is related to the DVC field of scan 007. The (small) force drop between scans 007 and 008 is due to the fact that the continuous acquisition between these two steps was interrupted by acquiring one high-quality scan, which is excluded from the present analyses.

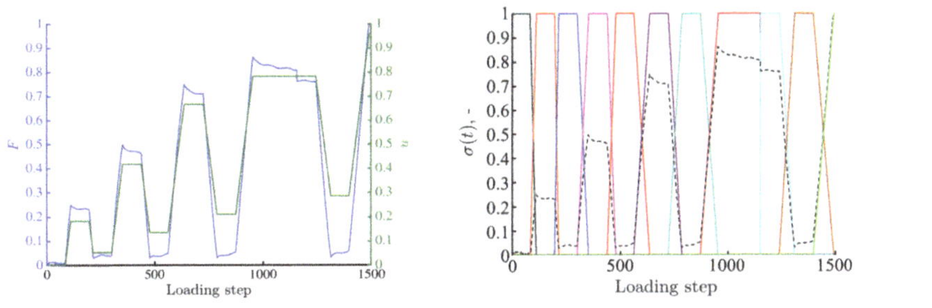

(**a**) (**b**)

Figure A5. (**a**) Temporal bases consisting of either normalized force history or stroke signal. (**b**) Temporal modes consisting of 11 plateau-like functions. The black dashed line represents the measured force signal normalized by its maximum value.

The initial value of the amplitude vector $\{\boldsymbol{\alpha}^0\}$ is set to 1 (Equation (7)). In all three cases, the P-DVC analysis takes four iterations to converge. Each iteration lasts approximately 5 h. The measured displacements yield a global decrease of the residuals within the investigated ROI, as shown in Figure A6. The rms projection residuals within the ROI are the lowest when using 11 plateau-like functions. From the reported results, it is shown that utilizing only one DVC displacement field (as a spatial mode) is sufficient to properly capture the kinematics up to specimen failure.

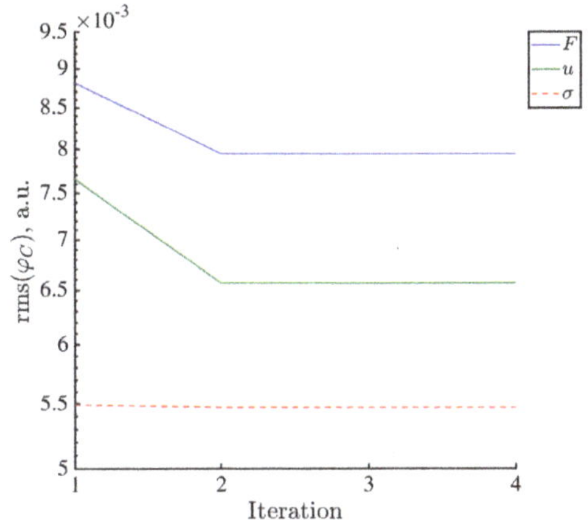

Figure A6. Change in rms projection residuals over the entire Region of Interest during the iterative procedure for different temporal interpolations.

Furthermore, the rms differences between the 3D nodal displacements obtained with FE-based DVC and projection-based measurements are calculated. Since large RBRs in x- and y-directions are observed during the experiment, these differences are only calculated for the z-direction displacements as the mechanical component is dominant in that direction. The rms differences are calculated between DVC displacement fields and converged P-DVC displacement fields for the time-steps corresponding to the beginning of each constant stroke stage. The obtained values are gathered in Table A3. As the projections acquired at the maximum load levels are affected by time-dependent motions (i.e., stress-relaxation phenomena), which influence the reconstruction of the full volume, differences are expected. From the values reported in Table A3, it is concluded that, when using measured force or prescribed stroke history as a temporal basis in combination with a single spatial mode, the rms differences are higher for the unloaded stages. This is especially pronounced when using the measured uniaxial force as the temporal interpolation basis, due to the fact that part of the deformation mechanisms is deactivated [34]. For the loaded stages, the differences are very small. When employing 11 plateau-like temporal functions, the rms differences are very small. The reported values confirm that, despite using only one spatial mode (i.e., the displacement field of scan 007) when performing 4D measurements, the kinematics is very well captured. In addition, employing the stroke history as the temporal interpolation better captured the temporal fluctuations (as the rms differences are smaller compared to those obtained when using the measured uniaxial force). However, the rms projection residuals within the ROI are the lowest when using 11 plateau-like functions. With such settings, the kinematics of the specimen is properly captured. This observation is also consistent with the truncation analysis where it was shown that the truncation error is significantly lower if more modes are kept (compared to the error if only one mode is used, see Figure A4).

Table A3. Comparison of rms differences in z direction between the 3D nodal displacements measured with FE-DVC and projection-based measurements. The comparison is performed for the converged P-DVC displacement fields for the angles at the beginning of each constant stroke stage.

		F	u	σ
Scan	**Mechanical State**	z	z	z
000	Unloaded	0.05	0.05	0.005
001	Loaded	0.81	0.05	0.007
002	Unloaded	0.35	0.06	0.01
003	Loaded	0.88	0.06	0.007
004	Unloaded	1.56	0.22	0.012
005	Loaded	0.73	0.08	0.002
006	Unloaded	2.77	0.3	0.011
007	Loaded	0.78	0.09	0.012
008	Loaded	0.76	0.08	0.011
009	Unloaded	3.76	0.25	0.011

Table A4 gathers the converged amplitudes. The amplitudes obtained with normalized force or stroke signals are very close. In both cases, the temporal amplitudes are greater than 1, as the DVC displacement field of the last loading cycle prior to specimen failure is used as spatial mode, while the 4D characterization is performed beyond this stage (i.e., up to specimen failure). Despite this extrapolation, the displacement fields for all analyzed radiographs converge and are consistent. When using 11 plateau-like functions, the amplitudes are close to 1 for all loading steps, except for the last one where the amplitude is greater than 1. Once again, this result stems from the fact that the DVC displacement field of scan 007 (Figure 2) is used as spatial component of the last mode, whereas 4D kinematic measurements are performed up to specimen failure (beyond scan 007). The displacement fields measured with P-DVC beyond scan 009 converge and are consistent. From the reported data, it is concluded that the kinematics is very well captured with the chosen temporal sampling, even though the displacement field becomes more complex with the increase in load level. Based on the reported data, 11 plateau-like functions constrained to 11 spatial modes obtained with FE-DVC are chosen as a suited temporal basis (Section 3).

Table A4. Temporal amplitudes. The * highlights an extrapolation, namely, when employing 11-plateau like functions as temporal basis, the temporal interpolation function describing the last loading stage prior to failure was constrained to the spatial mode of scan 007. When using the measured uniaxial force or stroke signals, only the displacement field of scan 007 is considered as the spatial mode.

		F	u	σ
Scan	**Mechanical State**	α	α	α
000	Unloaded			1.005
001	Loaded			1.007
002	Unloaded			0.983
003	Loaded			1.007
004	Unloaded			1.012
005	Loaded			1.002
006	Unloaded			0.998
007	Loaded			1.011
008	Loaded			1.006
009	Unloaded			0.996
007 *	Loaded	1.206	1.197	1.128

Appendix C. Validation of the Methodology for Quantification of Damage Growth

To validate the damage detection procedure (Section 4), an element within which damage does not occur is analyzed in the sequel. It corresponds to element #3, which is located at the specimen bottom surface (Figure A7).

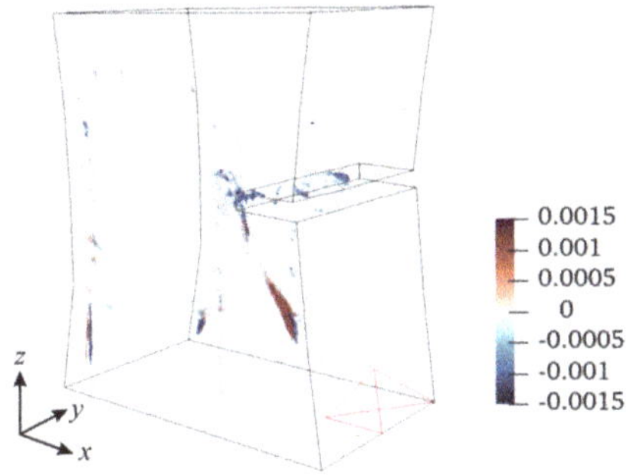

Figure A7. DVC residuals of scan 007. The red contour denotes element #3.

The DVC residuals within element #3 are projected onto the detector plane (Figure A8a,b). The first interval of residual increase for the loaded scans is within the range $\theta = 50°$ to 110°. The reported values are lower than those obtained within element #89 (Figure 7). There is an increase in correlation residuals with respect to levels of scan 000. However, it is not as pronounced as for element #89. It is also important to note that the residual levels obtained for the scans acquired in the loaded steps lie within the same range (Figure A8a). The values obtained for unloaded scans are within the range of scan 000 (Figure A8b). When plotting the rms DVC residuals of element #3 during the entire turns of scan acquisitions, the correlation residuals remain within the same range (Figure A8c). The projected residual map within element #3 calculated for scan 007 when $\theta = 90°$ is shown in Figure A8d. Their distribution is homogeneous, and the levels are low. It is concluded that damage did not occur within this element. This hypothesis will also be inspected with the projection residuals.

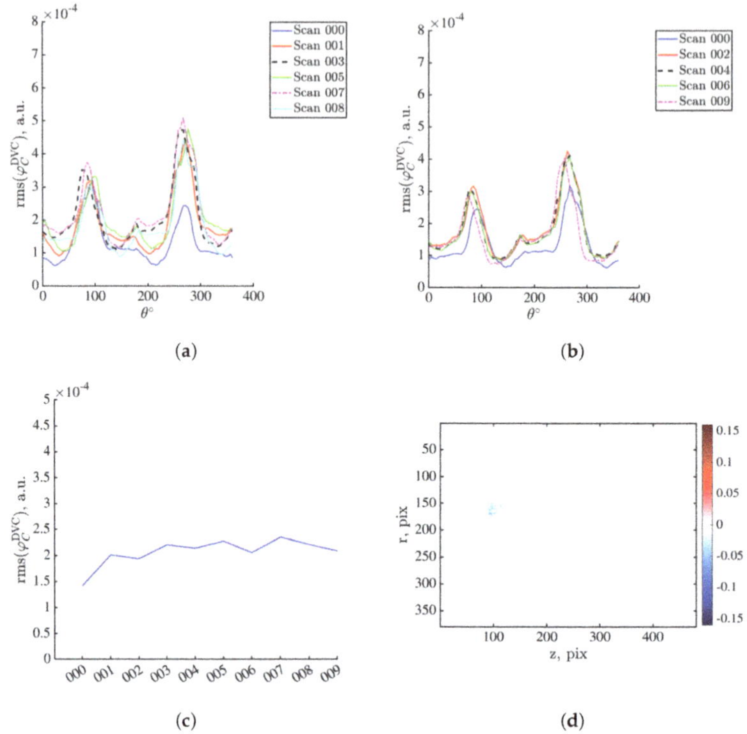

Figure A8. (**a**) Rms DVC residuals for element #3 projected onto the detector plane computed for the scans acquired at the peak of the loading cycles (Figure 2b). (**b**) Corresponding residuals computed for the scans acquired at the unloaded steps (Figure 2b). (**c**) Rms DVC residuals of element #3 during the entire turns of scan acquisitions. (**d**) Projected DVC residuals of element #3 when $\theta = 90°$.

As the first interval of DVC residual increase is within the range $\theta = 50°$ to $110°$ (Figure A8), the projection residuals are thus only inspected for specific angles, namely, $\theta = 0$, 42°, 60° and 90° acquired during the loading history. Figure A9 shows the amplification factor for these angles. For all observed angles, the values for the whole loading history are within the same range. There are no significant increases in amplification levels that may highlight damage growth, except for the very last part of the experiment.

The SVD procedure is also applied to the projection residuals for each analyzed angle ($\theta = 0°$, 42°, 60°, and 90°) as explained in Section 4. The singular values are shown in Figure A10. For all four analyzed angles, the first singular value is higher than the second one, thereby indicating a good separation between them. Thus, only the first temporal mode of each angle is analyzed.

Figure A11 shows the first temporal modes computed for $\theta = 0°$, 42°, 60°, and 90°. When $\theta = 42°$ and 90°, the levels are within the range of the unloaded stage (here marked with turn 1, corresponding to the range of scan 000). However, increases in levels compared to the first turn are observed when $\theta = 0°$ and 60°. This trend is due to the fact that these angles may be impacted by damage occurring in other zones/elements. Still, these increases are deemed not significant. The reported results demonstrate that, in this specific case, the damage analysis is highly dependent on the analyzed angles.

Last, the first spatial modes of the projection residuals computed for the chosen angles are shown in Figure A12. For all analyzed angles, the spatial distribution within element #3 is homogeneous.

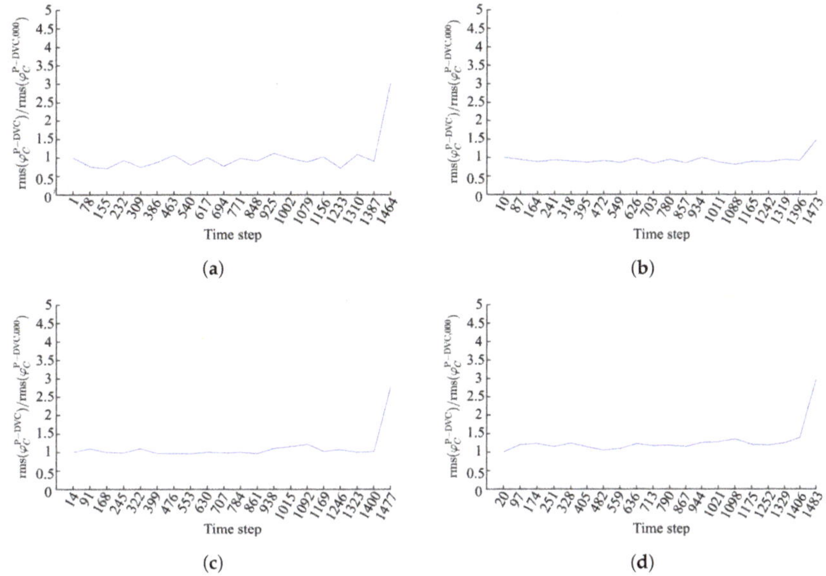

Figure A9. Amplification factor of projection residuals within element #3 when (a) $\theta = 0°$, (b) $\theta = 42°$, (c) $\theta = 60°$ and (d) $\theta = 90°$.

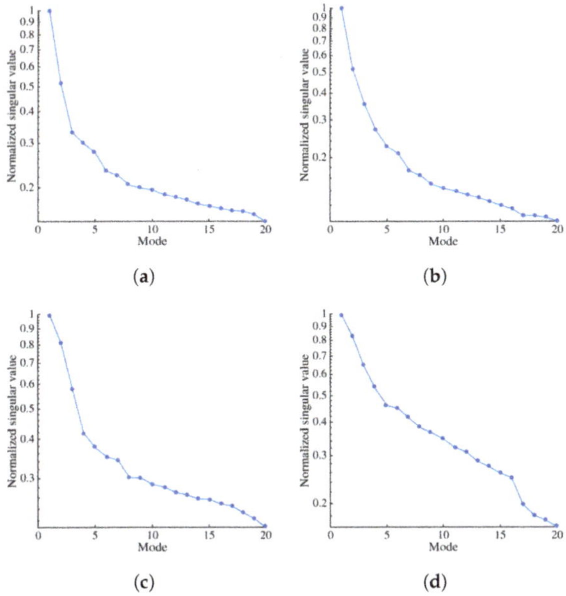

Figure A10. Normalized singular values for element #3 when (a) $\theta = 0°$, (b) $\theta = 42°$, (c) $\theta = 60°$, and (d) $\theta = 90°$.

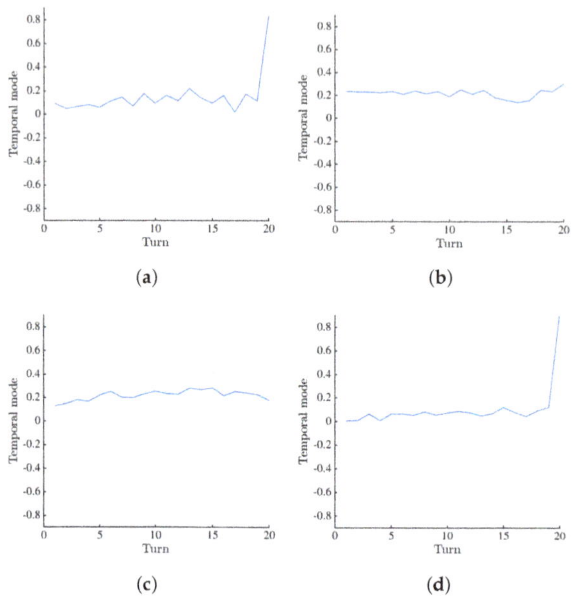

Figure A11. The first temporal mode of projection residuals within element #3 computed when (**a**) $\theta = 0°$, (**b**) $\theta = 42°$, (**c**) $\theta = 60°$, and (**d**) $\theta = 90°$.

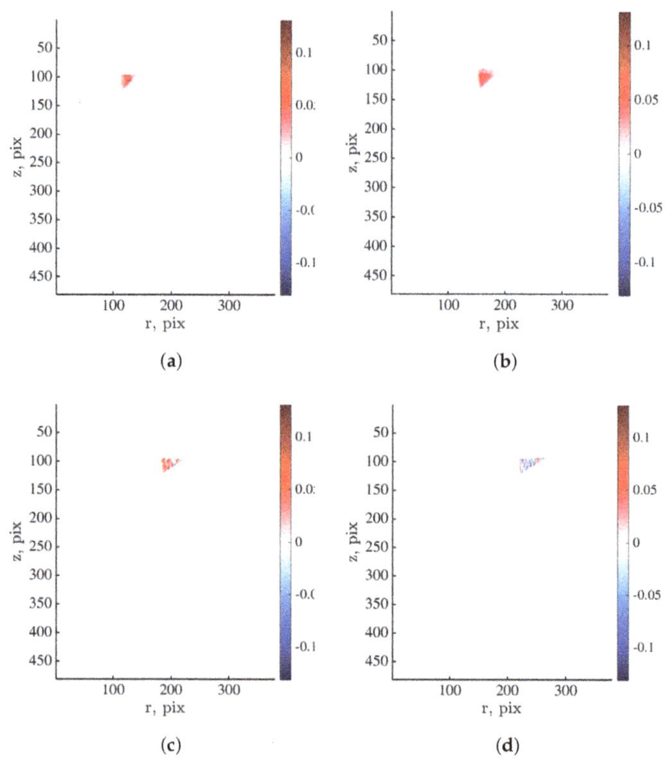

Figure A12. First spatial mode of the projection residuals within element #3 computed when (**a**) $\theta = 0°$, (**b**) $\theta = 42°$, (**c**) $\theta = 60°$, and (**d**) $\theta = 90°$.

References

1. Salvo, L.; Cloetens, P.; Maire, E.; Zabler, S.; Blandin, J.J.; Buffière, J.Y.; Ludwig, W.; Boller, E.; Bellet, D.; Josserond, C. X-ray micro-tomography an attractive characterisation technique in materials science. *Nucl. Instrum. Methods Phys. Res. Sect.-Beam Interact. Mater. At.* **2003**, *200*, 273–286. [CrossRef]
2. Maire, E.; Buffière, J.Y.; Salvo, L.; Blandin, J.J.; Ludwig, W.; Létang, J.M. On the application of x-ray microtomography in the field of materials science. *Adv. Eng. Mater.* **2001**, *3*, 539–546. [CrossRef]
3. Bay, B. Methods and applications of digital volume correlation. *J. Strain Anal.* **2008**, *43*, 745–760. [CrossRef]
4. Buljac, A.; Jailin, C.; Mendoza, A.; Neggers, J.; Taillandier-Thomas, T.; Bouterf, A.; Smaniotto, B.; Hild, F.; Roux, S. Digital Volume Correlation: Review of Progress and Challenges. *Exp. Mech.* **2018**, *58*, 661–708. [CrossRef]
5. Maire, E.; Bourlot, C.L.; Adrien, J.; Mortensen, A.; Mokso, R. 20 Hz X-ray tomography during an in situ tensile test. *Int. J. Fract.* **2016**, *200*, 3–12. [CrossRef]
6. García-Moreno, F.; Kamm, P.H.; Neu, T.R.; Bülk, F.; Mokso, R.; Schlepütz, C.M.; Stampanoni, M.F.M.; Banhart, J. Using X-ray tomoscopy to explore the dynamics of foaming metal. *Nat. Commun.* **2019**, *10*, 3762. [CrossRef]
7. Leclerc, H.; Roux, S.; Hild, F. Projection Savings in CT-based Digital Volume Correlation. *Exp. Mech.* **2015**, *55*, 275–287. [CrossRef]
8. Jailin, C.; Bouterf, A.; Poncelet, M.; Roux, S. In situ μ CT-scan Mechanical Tests: Fast 4D Mechanical Identification. *Exp. Mech.* **2017**, *57*, 1327–1340. [CrossRef]
9. Jailin, C.; Buljac, A.; Bouterf, A.; Hild, F.; Roux, S. Fast four-dimensional tensile test monitored via X-ray computed tomography: Single projection–based digital volume correlation dedicated to slender samples. *J. Strain Anal. Eng. Des.* **2018**, *53*, 473–484. [CrossRef]
10. Prashanth, S.; Km, S.; Nithin, K.; Sachhidan, S. Fiber Reinforced Composites—A Review. *J. Mater. Sci. Eng.* **2017**, *6*, 1–6.
11. Cosmi, F.; Ravalico, C. Threshold Identification for Micro-Tomographic Damage Characterisation in a Short-Fibre-Reinforced Polymer. *Strain* **2015**, *51*, 171–179. [CrossRef]
12. Sket, F.; Enfedaque, A.; Lopez, C.D.; González, C.; Molina-Aldareguia, J.M.; Llorca, J. X-ray computed tomography analysis of damage evolution in open hole carbon fiber-reinforced laminates subjected to in-plane shear. *Compos. Sci. Technol.* **2016**, *133*, 40–50. [CrossRef]
13. Rolland, H.; Saintier, N.; Wilson, P.; Merzeau, J.; Robert, G. In situ X-ray tomography investigation on damage mechanisms in short glass fibre reinforced thermoplastics: Effects of fibre orientation and relative humidity. *Compos. Part B-Eng.* **2017**, *109*, 170–186.
14. Ortiz-Morales, A.M.; Hanhan, I.; Solano, J.J.; Sangid, M.D. Observing progressive damage in carbon fiber epoxy laminate composites via 3D in situ X-ray tomography. *Eng. Fract. Mech.* **2021**, *246*, 107626. [CrossRef]
15. Buljac, A.; Navas, V.M.T.; Shakoor, M.; Bouterf, A.; Neggers, J.; Bernacki, M.; Bouchard, P.; Morgeneyer, T.; Hild, F. On the calibration of elastoplastic parameters at the microscale via X-ray microtomography and digital volume correlation for the simulation of ductile damage. *Eur. J. Mech. A-Solids* **2018**, *72*, 287–297. [CrossRef]
16. Tomičević, Z.; Bouterf, A.; Surma, R.; Hild, F. Damage observation in glass fiber reinforced composites via μ-tomography. *Mater. Today Proc.* **2019**, *12*, 185–191. [CrossRef]
17. Brault, R.; Germaneau, A.; Dupré, J.; Doumalin, P.; Mistou, S.; Fazzini, M. In-situ Analysis of Laminated Composite Materials by X-ray Micro-Computed Tomography and Digital Volume Correlation. *Exp. Mech.* **2013**, *53*, 1143–1151. [CrossRef]
18. Croom, B.; Jin, H.; Mills, B.E.; Carroll, J.; Long, K.; Brown, J.A.; Li, X. Damage mechanisms in elastomeric foam composites: Multiscale X-ray computed tomography and finite element analyses. *Compos. Sci. Technol.* **2019**, *169*, 195–202. [CrossRef]
19. Wang, K.; Pei, S.; Li, Y.; Li, J.; Zeng, D.; Su, X.; Xiao, X.; Chen, N. In-situ 3D fracture propagation of short carbon fiber reinforced polymer composites. *Compos. Sci. Technol.* **2019**, *182*, 107788.
20. Roux, S.; Hild, F.; Viot, P.; Bernard, D. Three dimensional image correlation from X-ray computed tomography of solid foam. *Compos. Part A-Appl. Sci. Manuf.* **2008**, *39*, 1253–1265. [CrossRef]
21. Réthoré, J.; Tinnes, J.; Roux, S.; Buffière, J.; Hild, F. Extended three-dimensional digital image correlation (X3D-DIC). *Comptes Rendus Mécanique* **2008**, *336*, 643–649. [CrossRef]
22. Leclerc, H.; Périé, J.; Roux, S.; Hild, F. Voxel-scale digital volume correlation. *Exp. Mech.* **2011**, *51*, 479–490. [CrossRef]
23. Réthoré, J.; Limodin, N.; Buffière, J.; Hild, F.; Ludwig, W.; Roux, S. Digital volume correlation analyses of synchrotron tomographic images. *J. Strain Anal. Eng. Des.* **2011**, *46*, 683–695. [CrossRef]
24. Tomičević, Z.; Kodvanj, J.; Hild, F. Characterization of the nonlinear behavior of nodular graphite cast iron via inverse identification. Analysis of uniaxial tests. *Eur. J. Mech.-A/Solids* **2016**, *59*, 330–343. [CrossRef]
25. Vrgoč, A.; Tomičević, Z.; Smaniotto, B.; Hild, F. Application of different imaging techniques for the characterization of damage in fiber reinforced polymer. *Compos. Part A Appl. Sci. Manuf.* **2021**, *150*, 106576. [CrossRef]
26. Maurer, J.; Jerabek, M.; Salaberger, D.; Thor, M.; Kastner, J.; Major, Z. Stress relaxation behaviour of glass fibre reinforced thermoplastic composites and its application to the design of interrupted in situ tensile tests for investigations by X-ray computed tomography. *Polym. Test.* **2022**, *109*, 107551. [CrossRef]
27. Vrgoč, A.; Tomičević, Z.; Smaniotto, B.; Hild, F. Characterization of glass fiber reinforced polymer via Digital Volume Correlation: Investigation of notch sensitivity. *Mech. Mater.* **2023**, *177*, 104552. [CrossRef]
28. Kosin, V.; Fau, A.; Jailin, C.; Smaniotto, B.; Wick, T.; Hild, F. A projection-based approach to extend digital volume correlation for 4D spacetime measurements. *Comptes Rendus Mécanique* **2023**, *351*, 265–280. [CrossRef]

29. Leclerc, H.; Neggers, J.; Mathieu, F.; Hild, F.; Roux, S. *Correli 3.0*; IDDN.FR.001.520008.000.S.P.2015.000.31500; Agence de Protection des Programmes: Paris, France, 2015.
30. van Aarle, W.; Palenstijn, W.J.; De Beenhouwer, J.; Altantzis, T.; Bals, S.; Batenburg, K.J.; Sijbers, J. The ASTRA Toolbox: A platform for advanced algorithm development in electron tomography. *Ultramicroscopy* **2015**, *157*, 35–47. [CrossRef]
31. Kak, A.; Slaney, M. *Principles of Computerized Tomographic Imaging*; IEEE Press: New York, NY, USA, 1988.
32. Jailin, C.; Buljac, A.; Bouterf, A.; Poncelet, M.; Hild, F.; Roux, S. Self-calibration for lab-μCT using space-time regularized projection-based DVC and model reduction. *Meas. Sci. Technol.* **2018**, *29*, 024003. [CrossRef]
33. Hallo, G.; Lacombe, C.; Parreault, R.; Roquin, N.; Donval, T.; Lamaignère, L.; Néauport, J.; Hild, F. Sub-pixel detection of laser-induced damage and its growth on fused silica optics using registration residuals. *Opt. Express* **2021**, *29 22*, 35820–35836. [CrossRef]
34. Vrgoč, A.; Tomičević, Z.; Smaniotto, B.; Hild, F. Characterization of glass fiber reinforced polymer via digital volume correlation: Quantification of strain activity and damage growth. *Compos. Sci. Technol.* **2023**, *234*, 109932. [CrossRef]

materials

MDPI

Article

Experimental Study of Drilling Damage Outcomes in Hybrid Composites with Waste Micro-Inclusions

Luis M. P. Durão [1,2,*], João E. Matos [2], João Alves [2], Sérgio Moni Ribeiro Filho [3], Túlio H. Panzera [3] and Fabrizio Scarpa [4]

1 Instituto de Ciência e Inovação em Engenharia Mecânica e Engenharia Industrial (INEGI), 4200-465 Porto, Portugal

2 Instituto Superior de Engenharia do Porto (ISEP), Instituto Politécnico do Porto (IPP), Rua Dr. António Bernardino de Almeida, 4249-015 Porto, Portugal; jem@isep.ipp.pt (J.E.M.); 1140629@isep.ipp.pt (J.A.)

3 Centre for Innovation and Technology in Composite Materials—CITeC, Department of Mechanical Engineering, Federal University of São João del Rei—UFSJ, São João del Rei 36301-360, MG, Brazil; sergiolmrf@gmail.com (S.M.R.F.); panzera@ufsj.edu.br (T.H.P.)

4 Bristol Composites Institute (ACCIS), University of Bristol, Bristol BS8 1TR, UK; f.scarpa@bristol.ac.uk

* Correspondence: lmd@isep.ipp.pt

Abstract: Composite materials are used in a substantial number of products. Environmental concerns highlight the need for the inclusion of recovered waste in their formulation, thus reducing their carbon footprint. These solutions raise the need to confirm the mechanical characteristics of these materials, avoiding unwanted failures. In this work, the authors present an experimental study on the drilling effects on fibrous–particulate hybrid composites made of glass/carbon fabrics and three different micro-inclusions: silica particles, recycled carbon fibre powder and cement. The mechanical features of the plates are confirmed by thrust force monitoring during drilling and by flexural testing. The range of results confirm the mechanical outcomes due to machining. The plates with monolithic carbon fabric or with carbon fabric plies in the outer plies returned higher mechanical characteristics. The plates with micro-inclusions had enhanced the flexural strength by 23% and 10%, in 40% and 60% fabric plates, respectively. The results demonstrate that the use of alternative formulations with micro-inclusions from recovered waste can contribute both to the reduction of the mechanical degradation of drilled hybrid composites and to environmental purposes by avoiding the increase in landfill waste.

Keywords: hybrid composites; waste reuse; drilling; flexural testing

Citation: Durão, L.M.P.; Matos, J.E.; Alves, J.; Filho, S.M.R.; Panzera, T.H.; Scarpa, F. Experimental Study of Drilling Damage Outcomes in Hybrid Composites with Waste Micro-Inclusions. *Materials* **2023**, *16*, 7325. https://doi.org/10.3390/ma16237325

Academic Editor: Shinichi Tashiro

Received: 31 October 2023
Revised: 17 November 2023
Accepted: 19 November 2023
Published: 24 November 2023

1. Introduction

Polymeric matrix composites, or fibre-reinforced polymers (FRP), have registered a considerable increase in their use, mainly since the second half of the past century. Their advantages are related to their low specific weight, good mechanical strength and stiffness. The continuous decrease in the price of materials and processing costs, together with the increase in reliability, have led to permanent innovation and to new applications. As the main advantages of composite materials lead to this extensive usage, a major drawback has arisen, which is related to the problem of recyclability at the end of their useful life. A possible alternative, as long as mechanical strength is not playing a major role, is the incorporation of particles in the formulation of composites.

FRPs are, by definition, a combination of two macro constituents—matrix and reinforcement—that are insoluble in each other [1]. Thus, in FRP parts the matrix is responsible for the bulk form of the part and chemical protection of the fibres, while the reinforcement is responsible for the mechanical resistance. Parts are designed in a way to take advantage of the fibre distribution to enhance mechanical resistance but, due to assembly or design purposes, they usually exhibit stress concentrations resulting from required assembly holes or imposed section changes. In the most common FRP parts only one matrix material and one reinforcement fibre are used. If there is more than one type of reinforcement, the composite is known as a hybrid. The need to address environmental issues, in addition to representing an option for recycling by-products from other industries, creates an opportunity for the incorporation of particles from discarded FRP parts as an alternative to the mere disposal of these parts in landfills [2]. Another foreseen advantage is the contribution to the overall demand for the reduction of carbon emissions. Different approaches are possible regarding these environmental concerns. In [3] Ribeiro Filho et al. presented a study on hybrid bio-composites reinforced with sisal–glass fibres and Portland cement particles, showing that the inclusion of cement particles had enhanced the flexural properties of the hybrid composites and that the stacking sequence strongly affected their mechanical properties. Gemi et al. [4] experimentally analysed the behaviour of pultruded GFRP composite beams infilled with hybrid fibre-reinforced concrete under four-point flexural loading. Furthermore, the authors analysed and then developed a numerical model to show the positive outcomes of this solution. In [2], Oliveira et al. presented a novel hybrid polymer composite made from sugarcane bagasse and discarded rubber particles, concluding that the mechanical and physical properties were substantially affected by the amount and size of the rubber particles and that the presence of sugarcane bagasse increased the compressive toughness of hybrid composites. The use of sustainable composites for transport applications had been presented in [5,6] regarding the development of alternative sustainable lightweight composites to reduce weight and costs in the automotive industry and in [7] for railways, considering the calculations of carbon emissions and energy consumption for the overall life cycle of sleepers, including recyclability.

Normally, parts are produced for supplementary grouping in structures; therefore, drilling is normally needed for enabling rapid assembly with rivets or bolts. One of the concerns when machining FRPs is related to the reinforcements, usually of an abrasive nature, causing rapid tool wear and deterioration of the machined surfaces. Typical damages after drilling are pushout delamination, peel-up delamination, fibre pull-out, burrs, splintering, swelling as well as some thermal damages due to matrix softening. When considering the drilling of composites, hole quality results are mainly fibre related [8]. The development of thrust forces during the drilling operation is accepted to be more related with the drill geometry, feed rate and spindle speed, as already demonstrated in several published studies by different authors, see [9–20], showing the importance of parameter selection for the minimization of machining damages. In [11] Khashaba suggested a new approach to the drilling of woven glass fibre-reinforced polymer (GFRP) composites: clamping the specimen with two support plates, thus controlling force at drill entry and exit sides. By using support plates, the thrust force was increased up to 8.3% compared to the unsupported conditions, but delaminations were decreased by about 65.5%. Gemi et al. [17,18] concluded that, regarding the stacking sequence, the one that considers a carbon layer between two glass layers (GCG) presented better performance in terms of mechanical properties and machinability characteristics. Particularly in [18], Gemi et al. were able to conclude that the stacking sequence considerably affects the damage modes and damage mechanisms of the composites. Moreover, lower delamination damage had occurred at lower feed rates and higher cutting speeds, as in the machining of composite plates. Moreover, a greater damage extension was formed in the GCG specimen. The drilling performance of stacked glass-carbon fibre-reinforced hybrid laminate composites was examined experimentally in [19]. In this work, Ergene et al. showed that the delamination can be decreased with higher cutting speeds or lower feed rates. A comprehensive study including the use of Taguchi techniques, analysis of variance (ANOVA) and the response surface method (RSM) seeking for cutting parameter optimization of hybrid fibre composite during drilling was conducted by Ozsoy et al. [20], concluding that a low cutting velocity and a low feed rate should be applied under dry conditions to reduce delamination. Alternative drilling strategies for delamination reduction have been studied. The use of cryogenic machining can reduce heat-generated damage and delamination [21] and the use of chilled air can also contribute to a decrease in delamination extension [22]. A different approach is suggested in [23], combining ultrasonic vibration and drilling to achieve higher hole quality.

The existence of a critical thrust force, F_{crit}, for the onset of delamination, according to the model presented by Hocheng and Dharan [24], is well known, and shown in Equation (1):

$$F_{crit} = \pi \frac{\sqrt{8G_{IC}E_1h^3}}{3\left(1 - v_{12}^2\right)} \quad (1)$$

This critical thrust force is a function of the elastic modulus, E_1, the Poisson ratio, v_{12}, the interlaminar fracture toughness in mode I, G_{Ic} and the uncut plate thickness, h.

Concerns regarding the outcome of hybrid composites, namely with glass and carbon fibre as reinforcement, were studied by different authors either under tensile, compressive [25] or flexural loading [26,27], showing that hybridisation strategies can be helpful to balance costs, performance and reliability of composites for structural and lightweight applications.

The main goal of the present work is to demonstrate that waste particles can be used as reinforcement in hybrid composites, avoiding their disposal in landfills, and thus contributing to environmental purposes. The experimental array presented here is an addition to a previous study on ultrasonic pulse velocity and physical properties of the same set of hybrid composites [28]. By using the same planning regarding the stacking sequence, volume fraction and type of particle, this study concerns the outcomes of the drilling operation on the mechanical properties of hybrid composites by flexural testing. Reviews on the problem of delamination onset, assessment and suppression [29–31] were considered. In [29], Liu et al. summarised the most recent progresses in the mechanical drilling of composite laminates, covering drilling operations, drill bit geometry and materials, drilling-induced delamination and its suppressing approaches, thrust force and tool wear. Kumar and Singh [30] focused on the issues related to the conventional and unconventional machining of composite materials, more specifically on drilling. Most recently, a general review of drilling-induced delamination for composite laminates, including a delamination formation mechanism, delamination quantification methodologies and measurement technologies and delamination suppression strategies was presented by Geng et al. [31]. These reviews have helped to obtain a comprehensive view on mechanical drilling of composite laminates and on the planning of the experimental work.

2. Materials and Methods

The experimental work aims to show the outcomes of drilling in hybrid composite plates made of glass/carbon fabrics and three diverse micro-inclusions like silica particles, recycled carbon microfibres and cement. See Tables 1 and 2 for the factors considered, experimental levels and plate identification.

Table 1. Matrix planning: factors and experimental levels.

Factors	Experimental Levels	Short Designation
Stacking sequence	Glass fibres only	G5
	Carbon fibres only	C5
	Glass/Glass/Glass/Carbon/Carbon	G3C2
	Carbon/Carbon/Glass/Glass/Glass	C2G3
	Carbon/Glass/Glass/Glass/Carbon	CG3C
	Glass/Carbon/Glass/Carbon/Glass	GCGCG
Volume fraction (wt%)	40/60 60/40	-
Particle type	No particle	No
	Silica, 9% wt, 2.7 g/cm^3	Si
	Cement, 9% wt, 2.8 g/cm^3	Ce
	Recycled carbon microfibres, 9% wt, 1.7 g/cm^3	CMF

Table 2. Experimental conditions and subsets.

Subset	Plate #	%Matr/Reinf (vol)	Particle Type	Stacking Sequence
S1	1	40/60	No	G5
	2	40/60	No	C5
	3	40/60	No	G3C2
	4	40/60	No	C2G3
	5	40/60	No	GCGCG
	6	40/60	No	CG3C
S2	7	40/60	Si	G5
	8	40/60	Si	C5
	9	40/60	Si	G3C2
	10	40/60	Si	C2G3
	11	40/60	Si	GCGCG
	12	40/60	Si	CG3C
S3	13	40/60	Ce	G5
	14	40/60	Ce	C5
	15	40/60	Ce	G3C2
	16	40/60	Ce	C2G3
	17	40/60	Ce	GCGCG
	18	40/60	Ce	CG3C
S4	19	40/60	CMF	G5
	20	40/60	CMF	C5
	21	40/60	CMF	G3C2
	22	40/60	CMF	C2G3
	23	40/60	CMF	GCGCG
	24	40/60	CMF	CG3C
S5	25	60/40	No	G5
	26	60/40	No	C5
	27	60/40	No	G3C2
	28	60/40	No	C2G3
	29	60/40	No	GCGCG
	30	60/40	No	CG3C
S6	31	60/40	Si	G5
	32	60/40	Si	C5
	33	60/40	Si	G3C2
	34	60/40	Si	C2G3
	35	60/40	Si	GCGCG
	36	60/40	Si	CG3C
S7	37	60/40	Ce	G5
	38	60/40	Ce	C5
	39	60/40	Ce	G3C2
	40	60/40	Ce	C2G3

Table 2. *Cont.*

Subset	Plate #	%Matr/Reinf (vol)	Particle Type	Stacking Sequence
	41	60/40	Ce	GCGCG
	42	60/40	Ce	CG3C
S8	43	60/40	CMF	G5
	44	60/40	CMF	C5
	45	60/40	CMF	G3C2
	46	60/40	CMF	C2G3
	47	60/40	CMF	GCGCG
	48	60/40	CMF	CG3C

So, 2 plates for each formulation were produced, resulting in a total amount of 96 plates, with the dimensions of $100 \times 100 \times 1\ mm^3$, prepared at the laboratories of Universidade Federal de São João del-Rei (UFSJR) in Brazil.

All the produced hybrid composites had 5 plies with different stacking sequences, 2 levels of reinforcement fraction and 4 types of particle inclusion (9% wt)—see Tables 1 and 2. Plates were cured for 12 h under 0.8 kPa and 24 °C followed by a stage of 7 days at room temperature. The final thickness was 1 mm. In Figure 2, SEM images of some plates are presented, showing the diverse morphology that results from different particle inclusion. More details on the plates can be found in Ribeiro Filho et al. [28].

These combinations resulted in a total of 48 different conditions, grouped in 8 subsets of 6 conditions, to enable the simplest comparison of the results. These diverse subsets, divided according to the reinforcement volume fraction and particle type, are referred to as S1 to S8 (see also Table 2). The target of reusing production scrap or waste to contribute to a reduction of the amount of material that will end in landfills, thus addressing environmental issues, is a line of work that has been followed by the authors. As already discussed, this experimental work follows a previous characterization of mechanical and physical properties and their correlation to ultrasonic pulse velocity and physical properties. Further information regarding plate manufacturing, including physical properties as well as a description of the raw materials and a complete DOE and SEM analysis can be found in [28].

Each of the 96 test plates was drilled according to the scheme shown in Figure 1, being the central section, with 6 holes, for the monitoring of thrust forces during drilling and the lateral holes to produce coupons for the flexural tests. The drilling sequence and subsequent plate cutting resulted in a total of 288 coupons (Figure 1).

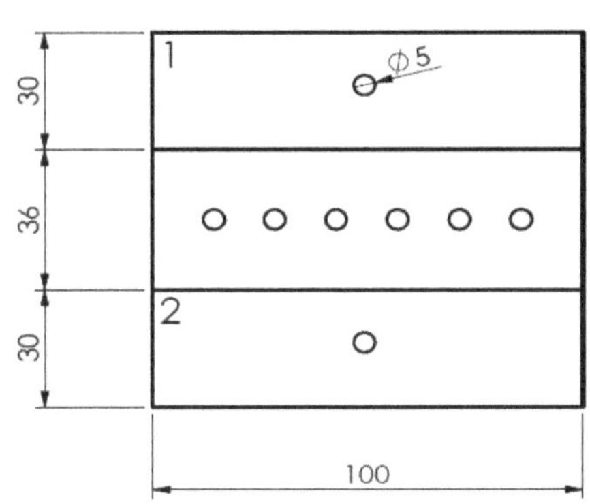

Figure 1. Drilling scheme of the plates. Plates numbered 1, 2 for tracking purpose.

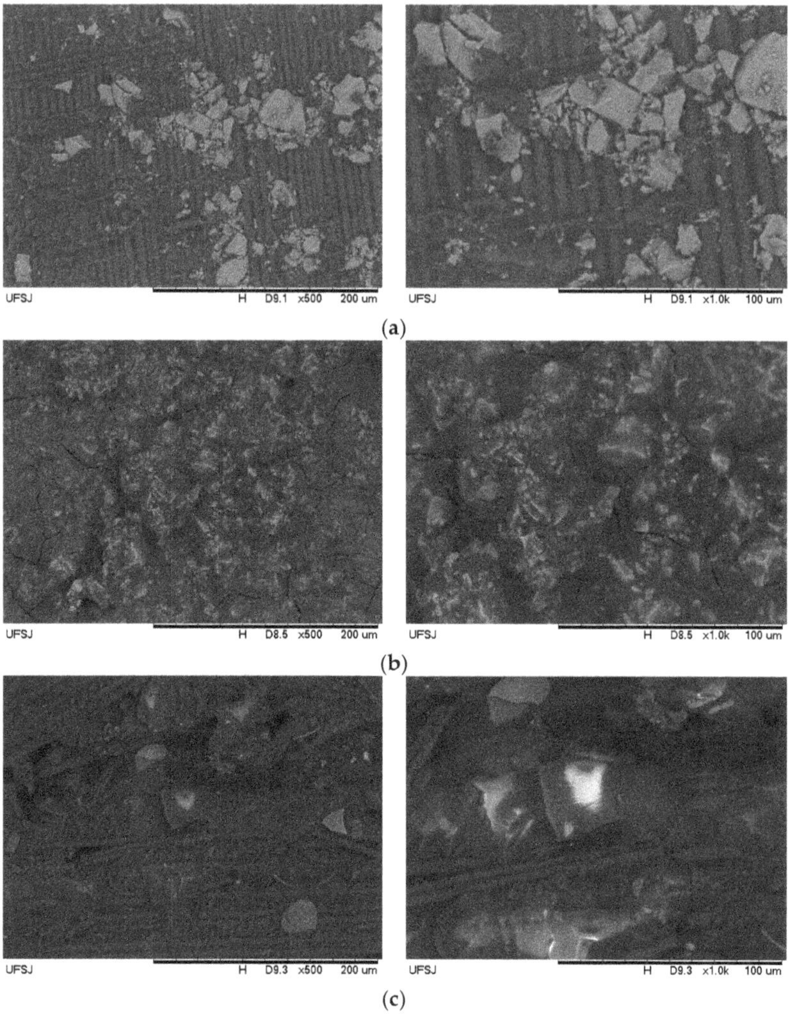

Figure 2. SEM images of some plates with fillers: (**a**) Silica (Plate 11); (**b**) Cement (Plate 17); (**c**) Carbon microfibres (Plate 23).

Drilling was performed on a HAAS VF-2 CNC and to enable a quick and steady positioning of the plates, a support was designed (see Figure 3 for experimental setup and plate support plate). The drilling parameters were set to 1120 rpm and a feed rate of 0.05 mm/rev for all the 5 mm diameter holes, performed with the help of a 5 mm diameter K20 tungsten carbide Brad point drill (see Figure 4) according to some of the papers referred above and recommendations of tool manufacturer. This drill geometry, due to its particular geometry, causes the tensioning of the fibres prior to the cut, enabling a clean surface. During the drilling operation, the thrust forces and torque were continuously monitored by using a KISTLER (Winterthur, Switzerland) 9171A rotating 4-component cutting force dynamometer associated with a computer for data collection. To avoid wear effects, the tool was replaced after every 4 plates were drilled.

Figure 3. Experimental setup and support for plate positioning.

Figure 4. Brad point drill.

The results considered as pertinent for this study were the thrust force during drilling and the flexural resistance of the drilled plates. The 4-point flexural test was performed at the Technological Testing Laboratory (LET) in ISEP, as described in the next section. For a complete and unbiased comparison of the results, a 3-point flexural test was performed on a similar set of non-drilled plates in UFJSR. The goal from the authors was to confirm equivalent results and conclusions from an experimental study carried out on carbon/epoxy plates [32]. After drilling, and before mechanical testing, the use of an NDT method such as enhanced radiography or another method was extensively referred to, in order to assess the delamination extension [33–35]. In the case of this work, NDT assessment was not performed due to the small thickness of the plates. So, it was considered that larger thrust forces would cause a larger delamination extension.

The final step of the experimental pipeline was the completion of a 4-point flexural test with the hole located at the centre of the coupons with a dimension of 100 × 30 × 1 mm^3 and a test speed (ν) equal to 9.2 mm/min, according to the conditions set in the ISO 14125:1998 standard [36], see Equation (2). In the 4-point bending flexural test there are no shear forces in the area between the two loading pins, only a constant bending moment. These flexural tests were carried out using a Shimadzu (Kyoto, Japan) AG-X Plus 100 kN universal testing machine.

$$v = \frac{\varepsilon' \times L^2}{4.7 \times h} \tag{2}$$

where ε' is a strain rate of 0.01 (or 1% per minute), L (distance between the supports) is a span equal to 66 mm and h the thickness of the coupon, equal to 1 mm.

The same standard was used for deriving the corresponding maximum flexural strength for each coupon, see Equation (3), considering a span of 66 mm (L) and a distance between the two arms of the anvil of 22 mm.

$$\sigma_f = \frac{F \times L}{b \times h^2} \tag{3}$$

where F is the load, in N, b is the coupon width, in mm and L and h are as in Equation (2).

As referred to above, a 3-point test was also performed on a set of similar plates, without drilling, in the laboratories of UFSJR, Brazil. These tests were performed using a Shimadzu (Kyoto, Japan) 100 kN universal testing machine, following ASTM D790:2017 [37], with a crosshead speed of 1 mm/min.

3. Results and Discussion

Considering the large number of coupons and respective results to present, the experimental results are summarised in Table 3, both for drilling thrust forces and 4-point flexural strength of the drilled plates and for the 3-point flexural test of undrilled plates. Then, for further discussion and analysis, the results are grouped in 8 subsets according to the following criteria: matrix/reinforcement volume fraction; type of particle; stacking sequence, as previously defined.

Table 3. Experimental results.

Subset	Plate #	Max Thrust (mN)	4P Flex Strgth [1] (MPa)	3P Flex Strgth [1] (MPa)
S1	1	2700	173	219
	2	3313	470	333
	3	2307	147	225
	4	2200	228	236
	5	3352	265	310
	6	3408	306	344
S2	7	1973	122	229
	8	3623	573	391
	9	2861	291	305
	10	3768	236	366
	11	3601	360	352
	12	3185	306	357
S3	13	2536	87	227
	14	4843	715	317
	15	3846	212	233
	16	3310	289	244
	17	3714	331	302
	18	3644	349	319
S4	19	2110	79	214
	20	3318	509	328
	21	3220	275	246
	22	3473	345	250
	23	3428	394	366
	24	3157	404	370
S5	25	2644	264	235
	26	4874	1031	480
	27	4303	466	478
	28	4021	426	402
	29	3553	396	396
	30	3722	353	235

Table 3. *Cont.*

Subset	Plate #	Max Thrust (mN)	4P Flex Strgth [1] (MPa)	3P Flex Strgth [1] (MPa)
S6	31	2989	241	295
	32	4973	1029	500
	33	2940	418	447
	34	3678	333	273
	35	3634	444	440
	36	3003	539	307
S7	37	2742	331	255
	38	4752	957	496
	39	3404	491	466
	40	3708	385	395
	41	3819	478	408
	42	3559	530	347
S8	43	2877	332	250
	44	4756	1036	504
	45	3563	534	406
	46	4150	467	381
	47	4318	548	396
	48	3440	596	336

[1] flex strgth = flexural strength.

3.1. Thrust Force during Drilling

The results of the thrust force monitoring were collected during the sequence of the six consecutive holes drilled in the central region of each plate (Figure 1). For comparison purposes, the maximum thrust force was the value considered, although the variation in this result was not significant along the different plates used in this experimental work. This outcome can be explained by the thickness of the plates, equal to 1 mm, and consequent low mechanical resistance to the drill cutting and breakthrough action as well as the option for conservative values of the drilling parameters.

The values presented in Table 3 are the average of the maximum thrust force registered for each condition of the plates (see Table 2 for identification). Due to the range of values observed during drilling, results are presented in mN instead of N. These values are then aggregated considering the grouping criteria in the subsets referred to above. For the fibre/reinforcement volume fraction subsets, an increase of 16% on the average of the maximum thrust force was noted when decreasing the volume fraction from 60% to 40%, as in the two left bars in Figure 5. This outcome can be related to the enhanced bulk effect of the matrix by assuring the mechanical link of the fibres and particles against the piercing action of the drill bit. Regarding the stacking sequence effect, the higher results of the thrust force were obtained for the C5 plates (carbon reinforced), as expected, with an increase of around 70% when compared with G5 plates (see the right-side bars in Figure 5). Interesting to mention is the effect of hybrid composites, combining two layers reinforced with carbon and three layers reinforced with glass (G3C2 or C2G3). In this case, there is a positive effect on the thrust force and a value that may be seen as a balanced one, combining lower costs with a reasonable mechanical strength, here evidenced by the thrust force. Particularly, the GCGCG plates, alternating glass-reinforced layers with carbon-reinforced layers seem promising, showing a maximum thrust force decrease of only 11% when compared to the C5 plates. These features can be compared with the results of the apparent density presented in [25], as this outcome is mainly dependent on the fibre arrangement and the slight increase in the apparent density caused by the particle inclusion, ranging from 1.7 to 2.8 g/cm^3, whereas the glass fibre or the carbon fibre densities are, respectively, 2.65 and 1.77 g/cm^3.

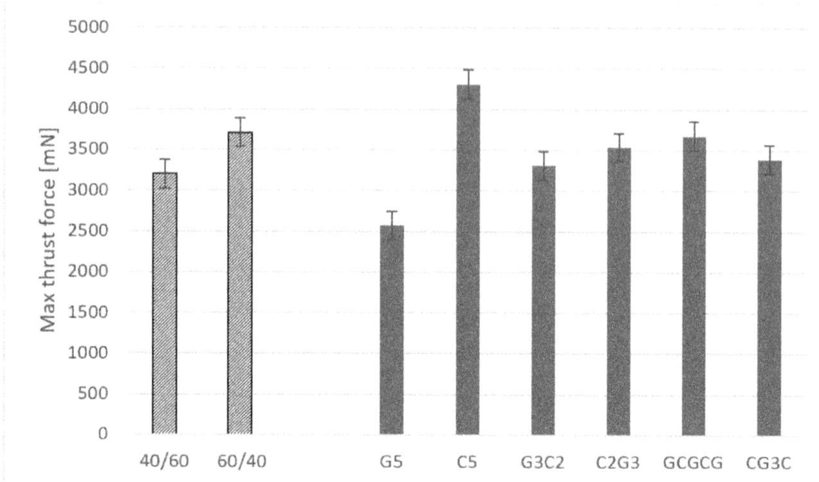

Figure 5. Maximum thrust force for different volume fractions and stacking sequences.

Finally, when considering the subsets S1 to S8, according to the particle type as defined in Table 2 a similar outcome can be seen, as the decrease in volume fraction has the same effect of raising the maximum thrust force needed to drill the plates by 17% in a global average (see Figure 6). It has to be noted that, due to the small thickness of the plates, the rise and drop of the thrust forces during drilling occurs in a short time, approximately one second; therefore, the peak force is seen as a good result for comparison of the plate behaviour during the drilling operation. Adding particles heightened the thrust force, independently of the particle type, for the plates with 60% volume fraction. For the plates with 40% volume fraction, the effect of added particles was not noticeable. It is interesting to note that a higher thrust force value was obtained for the plates without particles. This outcome shows that there is, on average, a positive effect on the reduced use of reinforcement fibres and that the use of particles can compensate for, in a limited but positive way, this reduction. Thus, the use of waste particles can be helpful in the mechanical characteristics of the plate under controlled conditions. As can be observed in Table 3, there was a moderate increase in the thrust force, meaning more resistance of the drilled plates to the punching action of the drill, more pronounced on the 60/40 plates. This aspect deserves further analysis, which is beyond the scope of the present work.

3.2. Flexural Testing

The 4-point flexural tests were performed on the drilled plates, with a 5 mm hole at the geometric centre of the plate (see Figure 1) and the test coupons were numbered 1 and 2 for each of the 96 plates. According to the experimental conditions described in Table 2 and in Figure 1, four coupons of the same type were tested for each of the 48 formulations. The aggregated results considering the defined subsets are presented in Figures 7 and 8. For these tests, the goal was to observe the outcome of the drilled plates when loaded in bending, as the existence of a drilled hole causes a reduction in the flexural strength of the plates, independently of the damage caused by the machining operation [32]. Even though the results of the 3-point flexural test performed in Brazil are presented in Table 3, they are considered mainly for comparison purposes, as the discussion on the outcomes is focused on the 4-point flexural test. Note that the major trends of the results, as discussed below, are essentially identical. As expected, due the major differences between the two test procedures, some significant dissimilarities were observed in the results presented in Table 3. For 3-point flexural tests, the plates were not previously drilled, and the development of both stresses is different due to the test setup. In 3-point flexural testing the peak stress is at the coupon mid-point and stress is reduced elsewhere. In 4-point flexural testing, the peak stress is exerted along an extended region of the coupon, hence exposing a larger length of the material, namely between the two inner points of the load application.

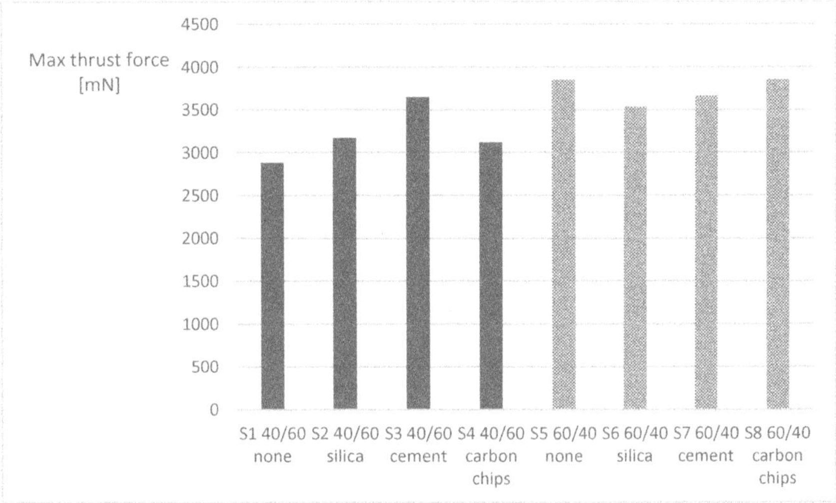

Figure 6. Maximum thrust force for sets S1 to S8 (as defined in Table 2).

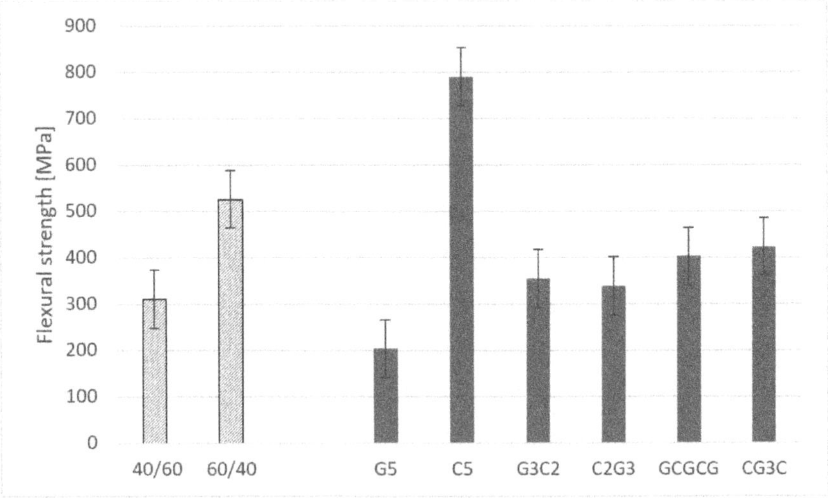

Figure 7. Flexural strength for different volume fractions and stacking sequences.

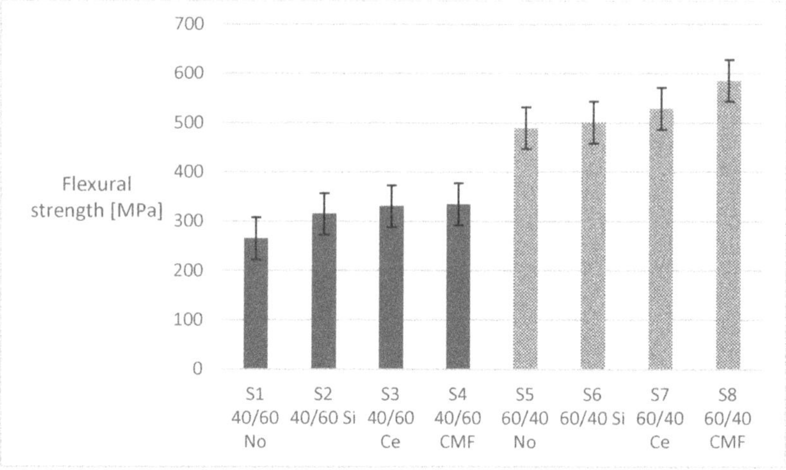

Figure 8. Flexural strength for sets S1 to S8 (as defined in Table 2).

As expected, the higher results in both test procedures were mostly observed when testing C5 plates, followed by the plates with carbon outer plies (G3C2, C2G3, CG3C). Some discrepancies are noteworthy when comparing 3-point with 4-point flexural test results, most of them on the plates with a 60/40% volume fraction with particles—plates 31 to 48. This outcome may be caused by some interaction of the particles with the reinforcement fibres, especially in non-drilled plates. For the subsets of 40/60% plates, the only odd result was verified when silica particles were added. These results will deserve further study to understand the possible interaction effects and their causes.

For the matrix/reinforcement volume fraction, the same bulk effect on the matrix/fibre content was observed, with an increase of about 70% on the flexural strength (see the left side bars in Figure 7). The most noteworthy effect is on the stacking sequence, as carbon plies have a definitive impact on the flexural strength, returning higher values for the plates that had carbon plies as outer plies (CGGGC)—see the rightmost bar in Figure 7—and a noteworthy improvement of almost 90% in comparison with the average values of the 4-point flexural strength for the C5 plates (carbon-reinforced only). This result can be explained by the superior mechanical strength of carbon fibres when compared to glass fibres. Therefore, the use of hybrid composites should bear in mind the positive effect of having carbon-reinforced outer plies if parts are supposed to bear bending loads. The inner plies reinforced with glass fibre do not show a negative effect on the mechanical features and make the hybrid composite less expensive. As already referred to in Section 3.1, these results have to be related to the outcomes of the apparent density presented in [25], concerning the effect of the stacking sequence and particle type on the density of composites, and thus on their physical properties.

Looking at the results of the 4-point flexural test ordered by the subsets S1 to S8, Figure 8, the higher results in flexural testing were observed for the plates with the higher matrix content, with an overall increase of 70%. This outcome is in accordance with the previous results presented in this work.

There is also an enhancing impact on the flexural strength when comparing the flexural strength of the plates with no particle with the plates with some particle added, independently of the particle type. On average, this increase can reach 20% in plates with less fibre content (40%). This increase is more evident when carbon microfibres were added. Further studies are needed to more deeply understand this effect on hybrid plates. This result highlights that the reuse of waste particles is possible, not causing any prejudicial effect on the mechanical properties of the composite plates.

4. Conclusions

In this work, hybrid composites reinforced with glass and carbon fabrics and three different micro-inclusions: silica particles, recycled carbon fibre powder and cement were assessed regarding their machinability and flexural resistance. The stacking sequence of glass and carbon plies was also varied, enabling a total of 48 different formulations. The following results from the experimental work can be presented.

The different ratios of matrix/reinforcement have consequences on the thrust force needed for drilling but also on the flexural strength of the hybrid plates.

The plates reinforced with carbon in the outer plies had higher flexural resistance, the C5 plates (carbon only) being the ones with the highest values, more than 20% above the values observed with other stacking sequences, on average.

These results can be related to the changes in the apparent density of the composites, due to the diverse densities of the reinforcement fibres.

Adding particles (9% weight), regardless of the type, increased the mechanical strength of the plates, as observed by the values of the drilling thrust force and flexural strength. For the drilling thrust force, however, this effect was less noted when the amount of reinforcement fibres rose from 40% to 60%. For the flexural strength there was a positive effect on the inclusion of particles, with the most noticeable result being when carbon microfibres were added. Again, these results are related to the slight change in the apparent density of composites with these micro-inclusions.

Looking closely at the effect of the inclusion of the different particles, some increase in the flexural strength was noted in some plates, but the main point is that these inclusions do not cause a decrease in the flexural strength, and thus can contribute to the use of less fibre, successfully replacing them in some content.

These formulations of hybrid composite materials, including carbon microfibres or other particles like silica or cement, can contribute to the reduction of waste, to environmental objectives and to the saving of raw materials like reinforcement fibres, enabling a circular economy life cycle for these composites.

Author Contributions: L.M.P.D.: conceptualization, methodology, formal analysis, writing—review and editing, project administration, funding acquisition; J.E.M.: methodology, software, data curation; J.A.: methodology, validation, investigation, resources, writing—original draft preparation; S.M.R.F.: investigation, methodology, writing—original draft preparation; T.H.P.: supervision, conceptualization, writing—review, visualization; F.S.: conceptualization, writing—review. All authors have read and agreed to the published version of the manuscript.

Funding: The authors would like to thank the financial support provided by Brazilian Research Agencies, CAPES (MSc scholarship), CNPq (PQ-309885/2019-1) and FAPEMIG (PPM-00075-17) and the Portuguese Agency FCT (Fundacão para a Ciência e a Tecnologia)—Portugal (Project UID/EMS/0615/2019).

Data Availability Statement: Derived data supporting the findings of this study are available from the corresponding author upon request.

Acknowledgments: The authors wish to thank the Mechanical Technology Laboratory and the Technological Testing Laboratory at ISEP for the use of the necessary equipment to accomplish his study.

Conflicts of Interest: The authors declare no conflict of interest.

References

1. Schwartz, M.M. *Composite Materials Handbook*; McGraw Hill: New York, NY, USA, 1992.
2. Oliveira, P.R.; Ribeiro Filho, S.L.M.; Panzera, T.H.; Christoforo, A.L.; Durão, L.M.D.; Scarpa, F. Hybrid polymer composites made of sugarcane bagasse fibres and disposed rubber particles. *Polym. Polym. Compos.* **2021**, *29*, S1280–S1293. [CrossRef]
3. Ribeiro Filho, S.L.M.; Oliveira, P.R.; Vieira, L.M.G.; Panzera, T.H.; Freire, R.T.S.; Scarpa, F. Hybrid bio-composites reinforced with sisal-glass fibres and Portland cement particles: A statistical approach. *Compos. Part B Eng.* **2018**, *149*, 58–66. [CrossRef]
4. Gemi, L.; Madenci, E.; Ozkılıç, Y.O. Experimental, analytical and numerical investigation of pultruded GFRP composite beams infilled with hybrid FRP reinforced concrete. *Eng. Struct.* **2021**, *244*, 112790. [CrossRef]

5. Dhakal, H.N.; Ismail, S.O. Chapter 7—Future outlooks and challenges of sustainable lightweight composites. In *Sustainable Composites for Lightweight Applications*; Woodhead Publishing Series in Composites Science and Engineering; Elsevier: Amsterdam, The Netherlands, 2021; pp. 285–290.
6. Fantuzzi, N.; Bacciocchi, M.; Benedetti, D.; Agnelli, J. The use of sustainable composites for the manufacturing of electric cars. *Compos. Part C* **2021**, *4*, 100096. [CrossRef]
7. Kaewunruen, S.; Liao, P. Sustainability and recyclability of composite materials for railway turnout systems. *J. Clean. Prod.* **2021**, *285*, 124890. [CrossRef]
8. Hocheng, H.; Tsao, C.C. The path towards delamination-free drilling of composite materials. *J. Mater. Process. Technol.* **2005**, *167*, 251–264. [CrossRef]
9. Lopez-Arraiza, A.; Amenabar, I.; Sarrionandia, M.; Aurrekoetxea, J. Experimental Analysis of Drilling Damage in Biocomposite Laminates Manufactured by Resin Transfer Molding. *J. Biobased Mater. Bioenergy* **2011**, *5*, 483–490. [CrossRef]
10. Khashaba, U.A.; El-Sonbaty, I.A.; Selmy, A.I. Machinability analysis in drilling woven GFR/epoxy composites: Part I—Effect of machining parameters. *Compos. Part A* **2010**, *41*, 391–400. [CrossRef]
11. Khashaba, U.A. Mechanics of chip, delamination, and burr formation in drilling supported woven GFRP composites. *Alex. Eng. J.* **2023**, *79*, 181–195. [CrossRef]
12. Chaudhary, V.; Gohil, P.P. Investigations on drilling of bidirectional cotton polyester composite. *Mater. Manuf. Process.* **2016**, *31*, 960–968. [CrossRef]
13. Lotfi, A.; Li, H.; Dao, D.V. Machinability Analysis in Drilling Flax Fiber-Reinforced Polylactic Acid Bio-Composite Laminates. *Int. J. Chem. Mater. Biomol. Sci.* **2019**, *13*, 9.
14. Díaz-Álvarez, A.; Rubio-López, A.; Santiuste, C.; Miguélez, M.H. Experimental analysis of drilling induced damage in biocomposites. *Text. Res. J.* **2018**, *88*, 2544–2558. [CrossRef]
15. Díaz-Álvarez, A.; Díaz-Álvarez, J.; Santiuste, C.; Miguélez, M.H. Experimental and numerical analysis of the influence of drill point angle when drilling biocomposites. *Compos. Struct.* **2019**, *209*, 700–709. [CrossRef]
16. Belaadi, A.; Laouici, H.; Bourchak, M. Mechanical and drilling performance of short jute fiber-reinforced polymer biocomposites: Statistical approach. *Int. J. Adv. Manuf. Technol.* **2020**, *106*, 1989–2000. [CrossRef]
17. Gemi, L.; Koklü, U.; Yazman, S.; Morkavuk, S. The effects of stacking sequence on drilling machinability of filament wound hybrid composite pipes: Part-1 mechanical characterization and drilling tests. *Compos. Part B* **2020**, *186*, 107787. [CrossRef]
18. Gemi, L.; Morkavuk, S.; Köklü, U.; Yazman, S. The effects of stacking sequence on drilling machinability of filament wound hybrid composite pipes: Part-2 damage analysis and surface quality. *Compos. Struct.* **2020**, *235*, 111737. [CrossRef]
19. Ergene, B.; Bolat, C.; Karakilinc, U.; Burak Irez, A. A comprehensive investigation of drilling performance of anisotropic stacked glass-carbon fiber reinforced hybrid laminate composites. *Polym. Compos.* **2023**, *44*, 2656–2670. [CrossRef]
20. Ozsoy, N.; Eksi, S.; Ozsoy, M. Cutting parameters optimization of hybrid fiber composite during drilling. *Mater. Test.* **2023**, *65*, 291–302. [CrossRef]
21. Giasin, K.; Barouni, A.; Dhakal, H.N.; Featherson, C.; Zitoune, R.; Morkavuk, S.; Koklu, U. Microstructural investigation and hole quality evaluation in S2/FM94 glass-fibre composites under dry and cryogenic conditions. *J. Reinf. Plast. Compos.* **2021**, *40*, 273–293. [CrossRef]
22. Abish, J.; Samal, P.; Narenther, M.S.; Kannan, C.; Balan, A.S.S. Assessment of drilling-induced damage in CFRP under chilled air environment. *Mater. Manuf. Process.* **2018**, *33*, 1361–1368. [CrossRef]
23. Sun, Z.; Geng, D.; Meng, F.; Zhou, L.; Jiang, X.; Zhang, D. High performance drilling of T800 CFRP composites by combining ultrasonic vibration and optimized drill structure. *Ultrasonics* **2023**, *134*, 107097. [CrossRef] [PubMed]
24. Hocheng, H.; Dharan, C.K.H. Delamination during Drilling in Composite Laminates. *J. Eng. Ind.* **1990**, *112*, 236–239. [CrossRef]
25. Rajpurohit, A.; Joannès, S.; Singery, V.; Sanial, P.; Laiarinandrasana, L. Hybrid Effect in In-Plane Loading of Carbon/Glass Fibre Based Inter- and Intraply Hybrid Composites. *J. Compos. Sci.* **2020**, *4*, 6. [CrossRef]
26. Xian, G.; Guo, R.; Li, C. Combined effects of sustained bending loading, water immersion and fiber hybrid mode on the mechanical properties of carbon/glass fiber reinforced polymer composite. *Compos. Struct.* **2022**, *281*, 115060. [CrossRef]
27. Lal, H.M.; Uthaman, A.; Li, C.; Xian, G.; Thomas, S. Combined effects of cyclic/sustained bending loading and water immersion on the interface shear strength of carbon/glass fiber reinforced polymer hybrid rods for bridge cable. *Constr. Build. Mater.* **2022**, *314*, 125587. [CrossRef]
28. Ribeiro Filho, S.L.M.; Thomas, C.; Durão, L.M.D.; Christoforo, A.L.; Bowen, C.; Scarpa, F.; Panzera, T.H. Ultrasonic pulse velocity and physical properties of hybrid composites: A statistical approach. *Hybrid Adv.* **2023**, *2*, 100024. [CrossRef]
29. Liu, D.F.; Tang, Y.J.; Cong, W.L. A review of mechanical drilling for composite laminates. *Compos. Struct.* **2012**, *94*, 1265–1279. [CrossRef]
30. Kumar, D.; Singh, K.K. An approach towards damage free machining of CFRP and GFRP composite material: A review. *Adv. Compos. Mater.* **2015**, *24*, 49–63. [CrossRef]
31. Geng, D.; Liu, Y.; Shao, Z.; Lu, Z.; Cai, J.; Li, X.; Jiang, X.; Zhang, D. Delamination formation, evaluation and suppression during drilling of composite laminates: A review. *Compos. Struct.* **2019**, *216*, 168–186. [CrossRef]
32. Silva, P.; Matos, J.E.; Durão, L.M.P. Analysis of damage outcome in the strength of polymer composite materials. *J. Compos. Mater.* **2019**, *53*, 547–560. [CrossRef]

33. Durão, L.M.P.; Tavares, J.M.R.S.; De Albuquerque, V.H.C.; Gonçalves, D.J.S. Damage evaluation of drilled carbon/epoxy laminates based on area assessment methods. *Compos. Struct.* **2013**, *96*, 576–583. [CrossRef]
34. Nagarajan, V.A.; Rajadurai, J.S.; Kumar, T.A. A digital image analysis to evaluate delamination factor for wind turbine composite laminate blade. *Compos. Part B* **2012**, *43*, 3153–3159. [CrossRef]
35. Devesa, L.F.S. Análise e Quantificação do dano em Materiais Compósitos de matriz Polimérica. Master's Thesis, ISEP, Porto, Portugal, 2020.
36. *ISO 14125:1998*; Fibre Reinforced Plastic Composites—Determination of flexural properties. ISO: Geneva, Switzerland, 1998.
37. *ASTM D790:2017*; Standard Test Methods for Flexural Properties of Unreinforced and Reinforced Plastics and Electrical Insulating Materials. ASTM: West Conshohocken, PA, USA, 2017.

materials

MDPI

Article

Acoustic Emission and K-S Metric Entropy as Methods to Analyze the Influence of Gamma-Aluminum Oxide Nanopowder on the Destruction Process of GFRP Composite Materials

Katarzyna Panasiuk *, Krzysztof Dudzik, Grzegorz Hajdukiewicz and Norbert Abramczyk

Faculty of Marine Engineering, Gdynia Maritime University, 81-225 Gdynia, Poland; k.dudzik@wm.umg.edu.pl (K.D.); g.hajdukiewicz@wm.umg.edu.pl (G.H.); n.abramczyk@wm.umg.edu.pl (N.A.)
* Correspondence: k.panasiuk@wm.umg.edu.pl; Tel.: +48-58-5586484

Abstract: Composites are materials that are widely used in industry, including yachting, railway and aviation. The properties of these materials can be modified by changing the type of reinforcement, the type of matrix, as well as the use of additives in the form of fillers and nanofillers that improve their mechanical or specific parameters. Due to the fact that these materials are often used for important structures, computational models using FEM tools may not be sufficient to determine the actual strength parameters, and what is more, to check them during operation. When designing structures made of composite materials, it is necessary to use high safety factors due to their behavior under several different types of loads, which is still difficult to determine precisely. This situation makes these structures much heavier and characterized by much higher strength properties than those that would actually be needed. In this article, the Kolmogorov-Sinai (K-S) metric entropy was used to determine the transition from the elastic to the viscoelastic state in GFRP (glass fiber reinforced polymer) composite materials without and with the addition of nanoaluminum, during a static tensile test. Additionally, the acoustic emission method was used during the research. This signal was further processed, and graphs were made of the number of events and the amplitude as a function of time. The obtained values were plotted on tensile graphs. The influence of the nano-filler on these parameters was also analyzed. The presented results show that it is possible to determine additional parameters affecting the strength of the structure for any composite materials.

Keywords: acoustic emission; K-S metric entropy; nano-additives; GFRP; tensile test

Citation: Panasiuk, K.; Dudzik, K.; Hajdukiewicz, G.; Abramczyk, N. Acoustic Emission and K-S Metric Entropy as Methods to Analyze the Influence of Gamma-Aluminum Oxide Nanopowder on the Destruction Process of GFRP Composite Materials. *Materials* **2023**, *16*, 7334. https://doi.org/10.3390/ma16237334

Academic Editors: Nuno Calçada Loureiro and Luís Miguel Pereira Durão

Received: 4 November 2023
Revised: 21 November 2023
Accepted: 23 November 2023
Published: 25 November 2023

1. Introduction

Due to their properties, composite materials are used as construction materials in many industries, from windmills, wagons or their parts and airplanes to yachts. Due to the easy modification of the strength properties of these materials, scientists are developing newer technologies and compounds to obtain even better properties. One such possibility is to combine laminates with additives in the form of nanofillers to create new materials called nanocomposites. Nanocomposites are materials in which at least one component has dimensions on the nanometric scale, which is called a nanofiller.

Nanofillers can be classified according to their chemical nature and type of physical structure, but most often they are classified according to the shape of the particles [1–3]. The attractiveness of nanocomposites results from the fact that the polymer matrix and the nanofiller interact with each other already at the molecular level. Thanks to this, a nanofiller with dimensions below 100 nm, added in a small amount to the matrix (usually a few percent), can significantly change the selected properties of the composite material. Research began with other aluminosilicates and substances with a lamellar structure, using a number of polymers as polymer matrices [4–7]. Nanocomposites with aluminosilicate fillers are currently used for the production of car engine details, in the aviation and space industries, etc. [8–10]. In article [11], composites based on epoxy resin/graphene oxide (GO) and epoxy resin/reduced graphene oxide (rGO) were tested in terms of thermo-mechanical properties, with emphasis on the influence of the chemical groups present on surfaces enriched with nano-additives. Studies have shown that the presence of oxidizing groups on GO contributes to the improvement of exfoliation, intercalation and distribution of GO sheets in composites compared to composites based on rGO. In study [12], metal-based nano-additives were incorporated into the PLA matrix, thus examining their influence on the surface properties of the antibacterial activity and mechanical properties of the PLA nano-additive film. The main aim of the research was to determine how the addition of nanoparticles to PLA during the extrusion process affects the chemical composition, morphology and wettability of the surface and its further impact on the antibacterial effectiveness and mechanical properties of PLA-NP. The article [13] focused on the influence of multi-wall carbon nanotubes on the mechanical and electrical properties of epoxy resins and epoxy composites. By using carbon nanotubes as polymer reinforcement, higher tensile strength values and higher deformation percentages were achieved. Most articles mainly focus on graphene nano-additives [14]. In study [15], the influence of the weight percentage of aluminum oxide nanoparticles (0.25–0.5–0.75–1 wt) and the particle size of 60 nm mixed with a commercial epoxy resin (Kemapoxy 150) was used. Bending, hardness, tensile, erosion, water absorption and TGA (thermogravimetry) tests were performed. The experimental results showed an improvement in mechanical behavior using 0.25 wt% particles for both flexural strength and wear resistance by 7 and 67% compared to pure epoxy. The research aimed to dope two nanometal oxides with high-density polyethylene (HDPE) using roller mixing and thermal pressing techniques. The weight fractions of the nanofillers ranged from 2.0% to 8.0% in the obtained composite. Interlayer shear strength (ILSS) was performed on a combination of nanofiller additives (Al_2O_3 and TiO_2) in the polymer composite. It was observed that the simultaneous presence of Al_2O_3 and TiO_2 nanofillers improves ILSS [16–18].

Due to the continuous search for materials that can improve the mechanical properties of composites [19–21], in this article it was decided to test composites based on glass fibers (GFRP) with the addition of the gamma-aluminum oxide nanopowder nanofiller. In order to more precisely visualize the results and determine the impact of the nanoadditive on the destruction process of composite materials, tools such as the acoustic emission method and K-S metric entropy were used for analysis [22].

The static tensile test of composite materials itself allows for the determination of parameters related to the strength of the materials, but it does not determine the changes that occur in the structure of these materials, such as, among others, in the case of metals, the yield point. The characteristics of the destruction process of composite materials during loading are still not known and adequately characterized. The use of acoustic emission signals during loading makes it possible to observe phenomena occurring in composite materials and, after a broader analysis, to also identify them. It is necessary to use appropriate parameters, such as the amplitude and number of events. Kolmogorov-Sinai metric entropy, based on deformation parameters, allows for determining the place where a change in the nature of deformation occurs (structural change). The combination of these two methods and appropriate correlation allows results to be obtained that can contribute to a completely different approach to the design of composite structures, not only using safety factors, but also real values.

2. Materials and Methods

Acoustic emission used simultaneously with a static tensile test, among many methods for detecting damage in composite materials, is very sensitive. It enables the study of deformation in the elastic or plastic range [23–25]. It allows for the monitoring and identification of damage, from the microscopic level to macroscopic changes, including the total destruction of the tested material [26–28]. The use of mechanical tests and acoustic emission allows for the detection and identification of individual destruction processes of composite materials, such as fiber breakage, matrix-reinforcement interface cracking and the cracking of the matrix. For this purpose, AE signals recorded during tests should be analyzed. The basic parameters describing these signals are amplitude, RMS, number of events, rise time, duration and energy [29,30]. The events that occur while loading the sample are impossible to observe, but they can be heard. Depending on the type of destruction process, these are, of course, different frequencies and different amplitudes. Taking into account, for example, the process of matrix destruction, we are dealing with brittle cracking that does not have large amplitude values. The process of the destruction of the fibers themselves is characterized by a larger number of events related to the cracking of both the fibers and the matrix as well as the much longer duration of these events. This is also related to the modulus of the elasticity of the fibers, which is much larger than the modulus of the elasticity of the matrix. The above conclusions are not based on theory or literature, but own research [31]. For more detailed damage monitoring, it is recommended to include an appropriate tool or signal in the analysis. Several studies conducted by the authors of this article [22,31–33] show that the use of the acoustic emission method combined with mechanical tests allows for obtaining much more information about composite materials than in the case of standard methods.

Additionally, in addition to using the acoustic emission method in the research, it was decided to use a statistical method based on Kolmogorov-Sinai (K-S) metric entropy. This method assumes that the qualitative changes occurring at the structural threshold separating the elastic state from the plastic state correspond to a specific measurement point. Energy dissipation occurs in the system, and the deterministic chaos of data associated with this phenomenon causes entropy variability [22]. The preparation of input data and the accuracy of measurement tools are extremely important when using this method [34]. The minimum value of metric entropy near the transition from the elastic to the plastic state is determined by the point separating the individual process states. The value of the stress corresponding to the transition of the tested material from the elastic to the plastic state is determined on the basis of the 'critical' point [35,36]. This method, together with its application based on research on composite materials and others, has been previously described in detail [22].

Polyester-glass laminates with the addition of a filler in the form of gamma-aluminum nanopowder were used for the tests. In order to reduce the risk of improper mixing of the resin and the nanofiller, the use of infusion methods and improper infiltration of resin into the materials, e.g., by stopping or a significant change in viscosity, the manual lamination method was used. Polimal 1094-AWTP polyester resin and glass mat with a weight of 450 g/m^2 were used. The matrix additives, such as recycled filler and nanofiller, were combined by physical mixing using a mixer. Two research materials were made: A0—without the addition of nanofiller, A2—with 2% nanofiller content. Table 1 shows the % content (by weight) of the combined ingredients.

Table 1. List of A0 and A2 composite ingredients.

Sample	Resin	Matrix	Nanoadditive
	%	%	%
A0	60	40	0
A2	60	38	2

The composite materials created in the manufacturing process were prepared for static tensile testing in accordance with the standard [37]. In order to obtain the appropriate shape of the specimens and minimize the influence of cutting temperature, the water-cutting method was used. 30 samples were cut from each material. Figure 1 shows the dimensions of the specimen, in accordance with the standard. Figure 2 shows a photo of the specimens prepared for the test.

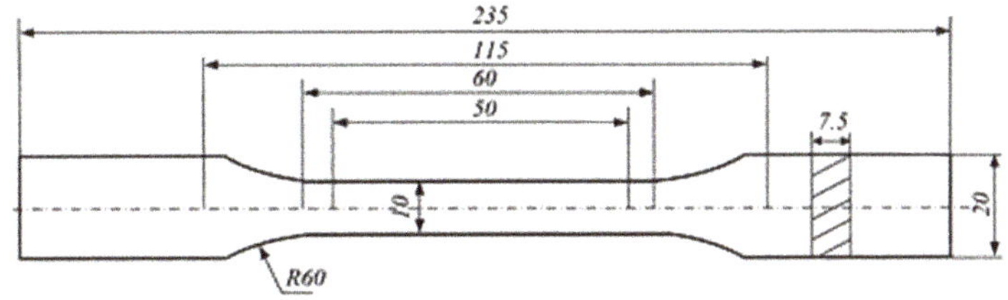

Figure 1. Dimensions of the specimens for the static tensile test in accordance with standard [31,37].

Figure 2. Selected specimens for testing.

The tests were carried out using AE measuring instruments (VallenSysteme GmbH, Icking, Munich, Germany) and a universal testing machine from Zwick Roel (Zwick Roell Group, Ulm, Germany). A piezoelectric sensor was installed on the tested specimens, which records the acoustic waves generated inside the material. These waves were converted by the sensor into an electrical signal, and then recorded in digital form by the recorder. The recorded signal was further processed and graphs of the effective value of the electrical signal (RMS), amplitude and hits as a function of time were plotted [22]. The instruments consisted of the AMSY-6 measurement system (VallenSysteme GmbH, Icking, (Munich), Germany); four ASIP-2 measurement (VallenSysteme GmbH, Icking, Munich, Germany) cards for recording and processing the AE signal; and VS-150-M sensors. AE research was performed using a set composed of a single channel recorder, USB AE Node, type 1283 with bandpass 20 kHz–1 MHz; a preamplifier with bandpass 75 kHz–1.1 MHz; AE-Sensor VS 150 M (with a frequency range of 100–450 kHz); and a computer with AE Win for USB Version E5.30 software for recording and analyzing AE data. The software used in the AE data acquisition was VisualAE (VallenSysteme GmbH, Icking, (Munich), Germany). Figure 3 shows a diagram of the measurement system [31].

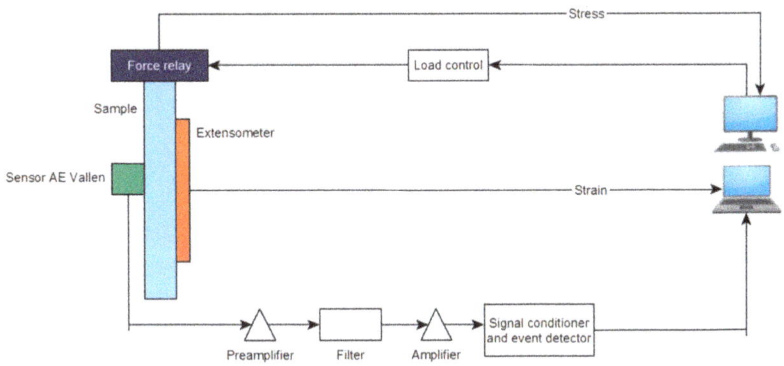

Figure 3. Diagram of measuring station [31].

In order to obtain accurate measurements and use the K-S metric entropy method, an extensometer was used to accurately measure the strain. The AE sensor was attached to the specimen using an elastic band, and a coupling fluid was used between the sensor and the specimen surface. Figure 4 shows the specimen with the AE sensor and extensometer.

Figure 4. Specimen being tested (2) with an AE sensor (1) and an extensometer (3).

3. Results and Discussion

During the static tensile test, 30 specimens from each material were tested. Table 2 presents a summary of the obtained results (average values) and Figure 5 shows graphs for two selected specimens.

Table 2. Summary of the results obtained from the static tensile test (30 specimens).

Sample	σ	E	ε
	MPa	**MPa**	%
A0	114.13	8927	1.71
Standard deviation	5.32	521	0.10
A2	128.26	9156	1.88
Standard deviation	8.40	434	0.12

Based on the results obtained from the static tensile test alone, it can be concluded that the nanofiller partially replacing the reinforcement improves the mechanical properties. As a result of the addition of the nanofiller, the tensile strength—σ—increased by about 11%, the elastic modulus—E—by about 3%, and the strain—ε—by about 9%. Despite the lack of chemical modifiers, the aluminum nanofiller combined with the resin did not adversely affect the strength parameters. It is worth mentioning that the addition of aluminum was 2% and it increased the strength properties by 9%. Therefore, it would be worth verifying in subsequent studies how larger amounts of the nanoadditive would affect the strength parameters.

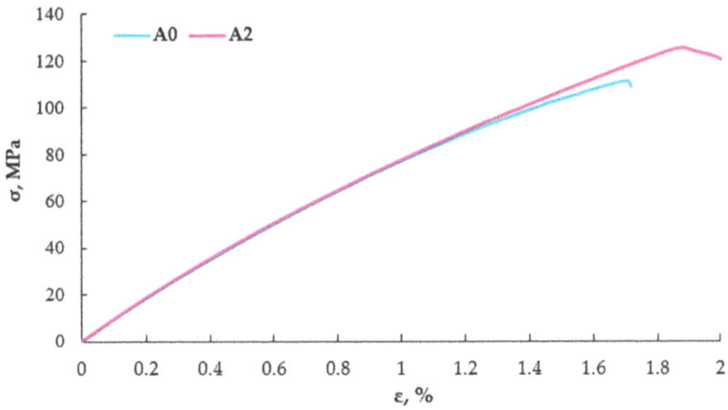

Figure 5. Tensile diagrams of selected specimens.

Based on previous studies [22,31–33], from among the many types of data obtained from acoustic emissions such as hits, RMS, amplitude, duration, etc., it was decided to select the parameter regarding the number of events and the analysis was based on the results of this parameter. Figures 6–8 show example charts obtained directly from the VisualAE software (Version 6.2).

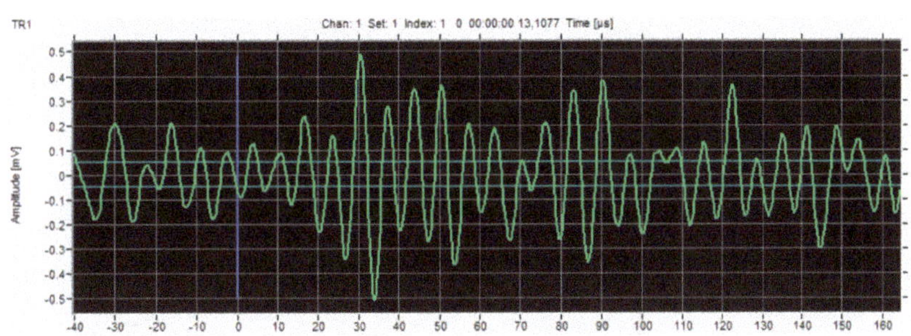

Figure 6. View of the signal amplitude change as a function of time—raw signal, without processing. (Green line is graph of an amplitude, blue line is a zero line).

FFT analysis is performed to find the characteristic value of the signal frequency for specific destruction mechanisms: matrix cracking, delamination, fiber cracking, etc. Thanks to preliminary research and subsequent comparison of these frequencies, it is possible to clearly determine the phenomenon generating a given signal.

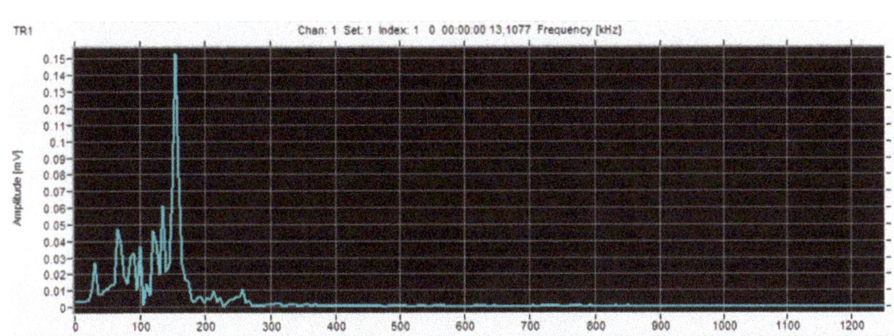

Figure 7. Changes in signal amplitude as a function of frequency—after FFT analysis.

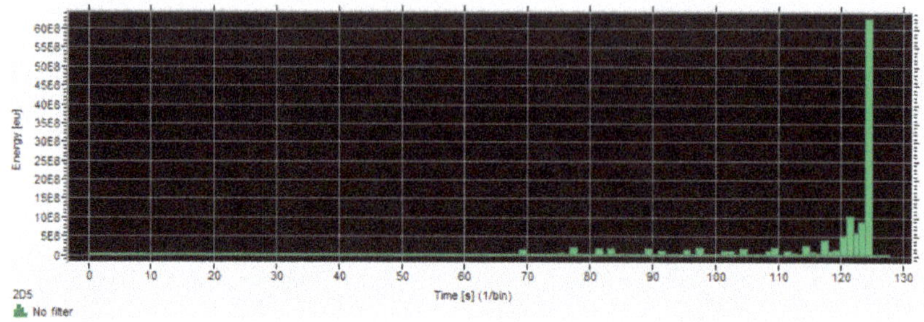

Figure 8. Change of signal energy over time.

The values obtained using the K-S metric entropy were also plotted on the obtained graphs. The parameters of strain, number of events (AE) and metric entropy are plotted as a function of time, since the K-S metric entropy is calculated based on the strain values obtained from the extensometer. In earlier work, the K-S metric entropy itself was presented in the charts as a function of measurement points, but for the purpose of comparing these results, the same reference point had to be adopted. Therefore, processing this data requires making at least three graphs to present the obtained values. Figure 9 shows the results obtained for an example sample without the addition of a nanofiller.

Figure 9a shows the values obtained from the acoustic emission as well as from the K-S metric entropy method. There is a noticeable common point for the decrease in entropy and the increase in the number of events at approximately the 76 s mark of the test. The previous values of the number of AE events were defined as the threshold because their values are comparable from the very beginning of the measurement. Except for one moment at 46 s, an additional graph was made to verify the amplitude value (Figure 10). The measurement using the acoustic emission method was started with a slight delay, hence the shift of the graph by a few seconds. Figure 9b shows the graphs of the number of AE events as a function of time, plotted on the strain ε (t) graphs, which made it possible to read the strain value at which this increase occurs. The strain for 76 s was approximately 1.1%. The same was done in the case of the K-S metric entropy (Figure 9c) and the same strain value was read from the graph. In Figure 9d, the value of the obtained strain for 76 s was plotted on the graph obtained from the static tensile test σ (ε), thanks to which the stress value was determined to be approximately 80 MPa. Earlier use of metric entropy in research, among others, aluminum alloy [36] made it possible to determine the structural change—the transition from the elastic to the plastic state in metals. In the case of composites, it is difficult to clearly define the point at this type of transition. In article [31], the emission results to determine the structural changes in the composite material were based on the amplitude, the values of which correspond to the number of events. Figure 10 shows the AE amplitude graph plotted on the K-S metric entropy graph.

The amplitude value at which the characteristic increase occurs (43 s) is approximately 47 dB. However, it can be omitted, taking into account the results obtained in previous studies and verified with those values that coincide with the K-S metric entropy (44 dB). In article [31], a detailed analysis of the destruction process using GFRP was carried out but with a different weight of the material used (350 g/m^2), which resulted in a reduction in the tensile strength of this material; however, by analyzing the results presented in Table 3, they can be compared with each other for the purpose of characterizing the point at 76 s, corresponding to approximately 1.1% deformation.

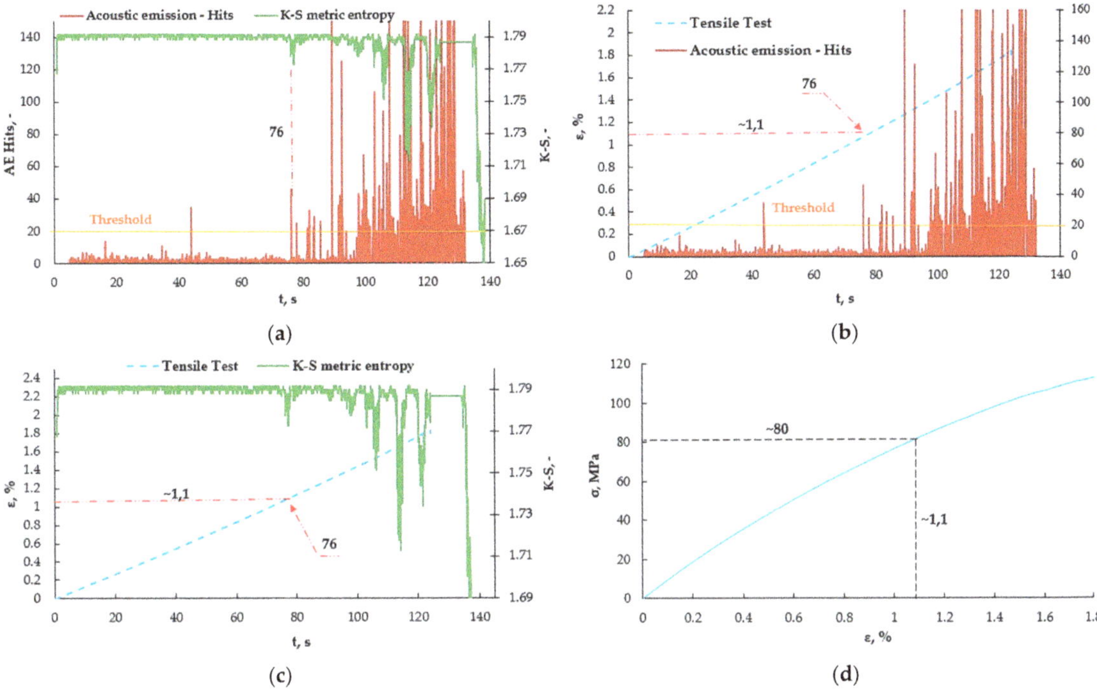

Figure 9. Results obtained from measurements using the acoustic emission method, using the K-S metric entropy for selected sample A0: (**a**) AE hits (t) plot plotted on the K-S (t) entropy graph, (**b**) graph ε (t) plotted on acoustic emission graph—AE hits (t), (**c**) ε (t) graph plotted on K-S (t) entropy graph, (**d**) obtained values plotted on a σ (ε) graph.

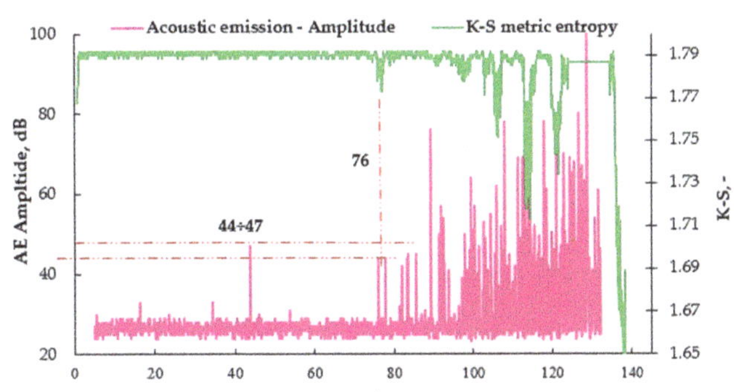

Figure 10. AE amplitude (t) graph plotted on the K-S (t) entropy graph for selected sample A0.

Table 3. Summary of the analysis of the degradation of the materials [31].

Signal Type	Amplitude	Stress
	dB	MPa
Matrix deformation	31–41	10–40
Cracking of matrix, delamination of fibers	42–50	38–50
Cracking of fibers	>50	51–110

Analyzing the results obtained for the amplitude at 76 s of the test, corresponding to 44 ÷ 47 dB, and the values in Table 3, it can be concluded that these two points indicate the beginning of the cracking of the matrix, with simultaneous fiber delamination. Hence, it can be characterized as the initiation of permanent and irreversible changes in the composite material. Figure 11 shows a similar analysis but using an aluminum nanoadditive.

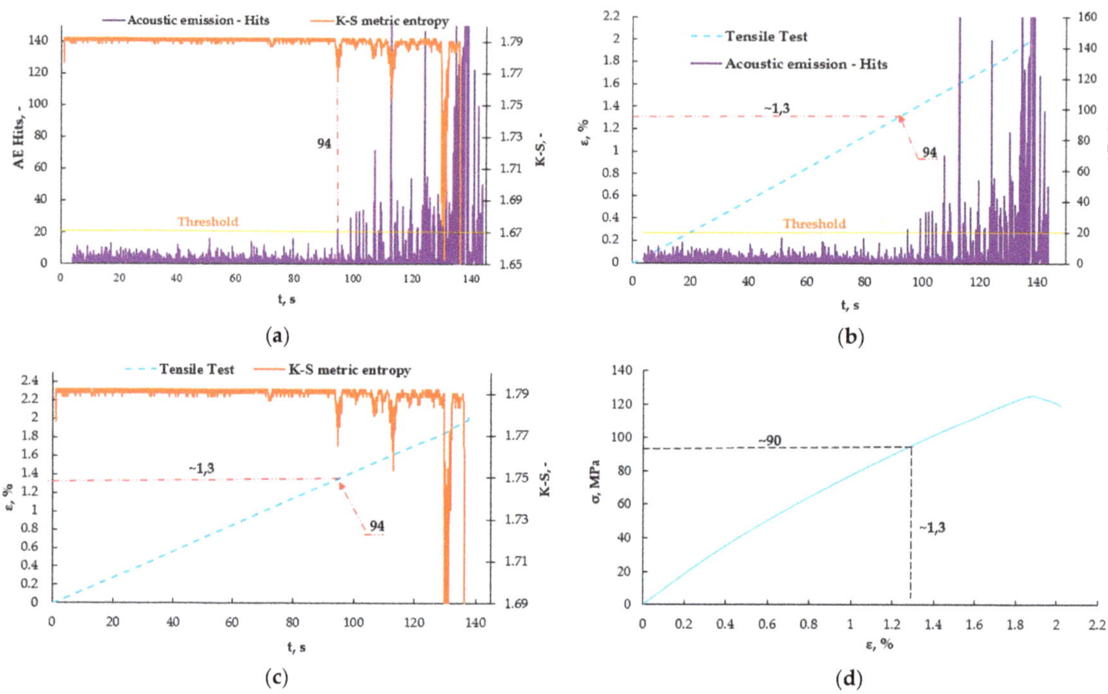

Figure 11. Results obtained from measurements using the acoustic emission method, using the K-S metric entropy for selected sample A2: (**a**) AE Hits (t) graph plotted on the K-S (t) entropy graph, (**b**) graph ε (t) plotted on acoustic emission graph—AE Hits (t), (**c**) ε (t) graph plotted on K-S (t) entropy graph, (**d**) obtained values plotted on a σ (ε) graph.

Figure 11a shows the values obtained from the acoustic emission as well as from the K-S metric entropy method. There is a noticeable common point for the decrease in entropy and the increase in the number of events at approximately the 94 s mark of the test. Figure 11b shows graphs of the number of AE events as a function of time, plotted on strain-rate—ε (t) graphs, which made it possible to read the strain value at which this increase occurs. The strain for 86 s was approximately 1.3%. The same was done for the K-S metric entropy (Figure 11c) and a strain value of approximately 1.3% was read from the graph. In Figure 11d, the value of the obtained strain for 86 s was plotted on the graph obtained from the static tensile test σ (ε), thanks to which the stress value was determined to be approximately 90 MPa. Figure 12 shows the AE amplitude graph plotted on the K-S metric entropy graph.

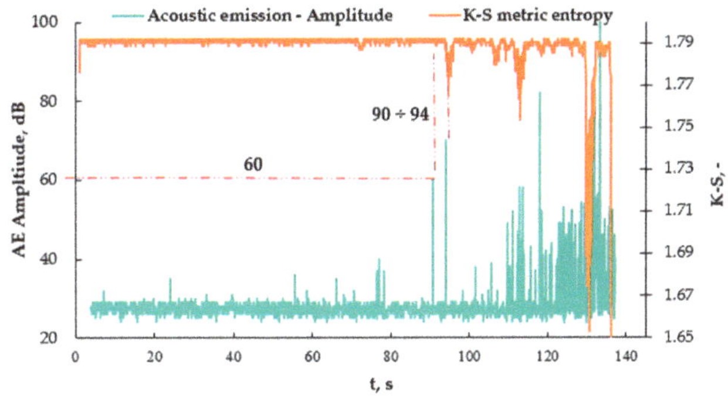

Figure 12. AE Amplitude (t) graph plotted on the K-S (t) entropy graph for selected sample A2.

Analyzing the graph (Figure 12) and the values in Table 3, it can be concluded that the addition of nanoaluminum to the resin significantly influenced the elasticity of the composite, extending the range of deformations for the matrix itself. It was increased from 80 MPa to 90 MPa, or by approximately 13%, by adding just 2% of the nanoadditive to the resin.

4. Conclusions

Nanoadditives are materials that are used to increase the strength parameters of composite materials. They may influence the variability of a number of parameters, such as tensile strength or strain. In this article, a 2% aluminum nanoadditive was used, thus replacing the reinforcement, in order to determine its impact on the strength parameters of the composite comparing the modified material to material without the additive. Using the K-S metric entropy method and the acoustic emission method, its influence on the mechanical and strength parameters was verified, not only on the basis of a static tensile test. The nanofiller, which partially replaces the reinforcement, improves the mechanical properties. As a result of the addition of the nanofiller, the tensile strength—σ—increased by about 11%, the elastic modulus—E—by about 3%, and the strain—ε—by about 9%. Despite the lack of the addition of chemical modifiers, the nanofiller had a positive effect on the mechanical properties. Using the K-S metric entropy method and the acoustic emission method, the values at which permanent deformations occur in the composite were determined, including the cracking of the matrix or delamination of the fibers. For the composite without the addition of aluminum, a value of 80 MPa was obtained, while thanks to the nano-addition of aluminum—an increase of 13% and a value of 90 MPa were obtained. Application of the abovementioned methods, as additional tests for new materials, allows for the determination of many more values that are particularly important when designing structures. Static tensile testing of composite materials, in most cases, provides a tensile strength value but does not involve yield stresses, commonly used to characterize metals. The composite material and its behavior during loading are not as known and characterized. The analysis of AE signals recorded during loading enables not only the observation of phenomena that occur during the material degradation processes, but also their identification. In laboratory tests, appropriate sensitivity thresholds should be determined for characteristic signal parameters, e.g., the amplitude and number of events. Knowing the values for materials that cause permanent changes allows for the determination of the permissible stress values for composites already during the design stage. This opens a new perspective on the research of composite materials and confirms the belief that there are research methods that allow for the verification of structural changes in composite materials, not only using tools at the micro-scale, such as SEM, but also at the macro scale.

Author Contributions: Conceptualization, K.P., K.D. and N.A.; data curation, K.D. and G.H.; formal analysis, K.P., K.D. and N.A.; methodology, K.P., K.D. and N.A.; resources, K.P.; software, K.D. and

G.H.; supervision, K.P.; visualization, K.P. and G.H.; writing—original draft, K.P.; writing—review and editing, K.P., K.D., G.H. and N.A. All authors have read and agreed to the published version of the manuscript.

Funding: This research received no external funding.

Institutional Review Board Statement: Not applicable.

Informed Consent Statement: Not applicable.

Data Availability Statement: Data available on request due to restrictions eg. privacy or ethical. The data presented in this study are available on request from the corresponding author. The data are not publicly available due to its huge amount.

Conflicts of Interest: The authors declare no conflict of interest.

References

1. Królikowski, W.; Rosłaniec, Z. Nanokompozyty polimerowe. *Composites* **2004**, *4*, 3–16.
2. Fusco, R.; Moretti, L.; Fiore, N.; D'Andrea, A. Behavior Evaluation of Bituminous Mixtures Reinforced with Nano-Sized Additives: A Review. *Sustainability* **2020**, *12*, 8044. [CrossRef]
3. Abramczyk, N.; Drewing, S.; Panasiuk, K.; Żuk, D. Application of Statistical Methods to Accurately Assess the Effect of Gamma Aluminum Oxide Nanopowder on the Hardness of Composite Materials with Polyester–Glass Recyclate. *Materials* **2022**, *15*, 5957. [CrossRef] [PubMed]
4. Spasówka, E.; Rudnik, E.; Kijeński, J. Biodegradowalne nanokompozyty polimerowe. *Polimery* **2006**, *51*, 617–626.
5. Nour, M.A. Polymer/clay nanocomposites. *Polimery* **2002**, *47*, 326–331. [CrossRef]
6. Yariv, S.; Cross, H. Introduction to Organo-Clay Complexes and Intercalations. *Organo-Clay Complexes Interact.* **2002**, *74*, 51–124.
7. Okamoto, M. *Polymer/Layered Silicate Nanocomposites*; Rapra Review Reports; RAPRA Technology: Shropshire, UK, 2003; Volume 14, p. 7.
8. Rabek, J.F. *Współczesna Wiedza o Polimerach*; Wydawnictwo Naukowe PWN: Warszawa, Poland, 2008.
9. Barton, J.; Niemczyk, A.; Czaja, K.; Korach, Ł.; Sacher-Majewska, B. Kompozyty, biokompozyty i nanokompozyty polimerowe. Otrzymywanie, skład, właściwości I kierunkizastosowań. *Chemik* **2014**, *68*, 280–287.
10. Yousri, O.M.; Abdellatif, M.H.; Bassioni, G. Effect of Al_2O_3 Nanoparticles on the Mechanical and Physical Properties of Epoxy Composite. *Arab. J. Sci. Eng.* **2018**, *43*, 1511–1517. [CrossRef]
11. Monteserín, C.; Blanco, M.; Aranzabe, E.; Aranzabe, A.; Laza, J.M.; Larrañaga-Varga, A.; Vilas, J.L. Effects of Graphene Oxide and Chemically-Reduced Graphene Oxide on the Dynamic Mechanical Properties of Epoxy Amine Composites. *Polymers* **2017**, *9*, 449. [CrossRef]
12. PušnikČrešnar, K.; Aulova, A.; Bikiaris, D.N.; Lambropoulou, D.; Kuzmič, K.; Fras Zemljič, L. Incorporation of Metal-Based Nanoadditives into the PLA Matrix: Effect of Surface Properties on Antibacterial Activity and Mechanical Performance of PLA Nanoadditive Films. *Molecules* **2021**, *26*, 4161. [CrossRef]
13. Smoleń, P.; Czujko, T.; Komorek, Z.; Grochala, D.; Rutkowska, A.; Osiewicz-Powęzka, M. Mechanical and Electrical Properties of Epoxy Composites Modified by Functionalized Multiwalled Carbon Nanotubes. *Materials* **2021**, *14*, 3325. [CrossRef] [PubMed]
14. Tarfaoui, M.; Lafdi, K.; Beloufa, I.; Daloia, D.; Muhsan, A. Effect of Graphene Nano-Additives on the Local Mechanical Behavior of Derived Polymer Nanocomposites. *Polymers* **2018**, *10*, 667. [CrossRef] [PubMed]
15. Mohamed, E.M.; Mohammed, A.K.; Rehab, M.; Mahues, R.Y. Effects of Al_2O_3 and Ba Onano-additives on mechanical characteristics of high-density polyethylene. *Mater. Chem. Phys.* **2021**, *262*, 124251. [CrossRef]
16. Nayak, R.K.; Mahato, K.K.; Routara, B.C.; Chandra Ray, B. Evaluation of mechanical properties of Al_2O_3 and TiO_2 nano filled enhanced glass fiber reinforced polymer composites. *J. Appl. Polym. Sci.* **2016**, *133*, 44274. [CrossRef]
17. Mahmoud, M.E.; El-Khatib, A.M.; El-Sharkawy, R.M.; Rashad, A.R.; Badawi, M.S.; Gepreel, M.A. Design and testing of high-density polyethylene nanocomposites filled with lead oxide micro- and nano-particles: Mechanical, thermal, and morphological properties. *J. Appl. Polym. Sci.* **2019**, *136*, 47812. [CrossRef]
18. Fu, S.-Y.; Feng, X.-Q.; Lauke, B.; Mai, Y.-W. Effects of particle size, particle/matrix interface adhesion and particle loading on mechanical properties of particulate–polymer composites. *Compos. Part B Eng.* **2008**, *39*, 933–961. [CrossRef]
19. Agureev, L.E.; Kostikov, V.I.; Eremeeva, Z.V.; Barmin, A.A.; Savushkina, S.V.; Ivanov, B.S. Aluminum Composites with Small Nanoparticles Additions: Corrosion Resistance. *Mech. Mater. Sci. Eng. J.* **2016**. [CrossRef]
20. Akl, S.; Abdel-Rehim, A.; Elsoudy, S. An Experimental Investigation of Tribological Performance of a Lubricant Using Different Nano Additives. *SAE Tech. Pap.* **2018**, *1*, 833.
21. Sławski, S.; Woźniak, A.; Bazan, P.; Mrówka, M. The Mechanical and Tribological Properties of Epoxy-Based Composites Filled with Manganese-Containing Waste. *Materials* **2022**, *15*, 1579. [CrossRef]
22. Kyzioł, L.; Panasiuk, K.; Hajdukiewicz, G.; Dudzik, K. Acoustic Emission and K-S Metric Entropy as Methods for Determining Mechanical Properties of Composite Materials. *Sensors* **2021**, *21*, 145. [CrossRef]

23. Šofer, M.; Cienciala, J.; Fusek, M.; Pavlíček, P.; Moravec, R. Damage Analysis of Composite CFRP Tubes Using Acoustic Emission Monitoring and Pattern Recognition Approach. *Materials* **2021**, *14*, 786. [CrossRef] [PubMed]
24. Zhuang, X.; Yan, X. Investigation of damage mechanisms in self-reinforced polyethylene composites by acoustic emission. *Compos. Sci. Technol.* **2006**, *66*, 444–449. [CrossRef]
25. Yu, Y.-H.; Cho, J.-H.; Kweon, J.-H.; Kim, D.-H. A study on the failure detection of composite materials using an acoustic emission. *Compos. Struct.* **2006**, *75*, 163–169. [CrossRef]
26. Saeedifar, M.; Najafabadi, M.A.; Zarouchas, D.; Toudeshky, H.H.; Jalalvand, M. Clustering of interlaminar and intralaminar damages in laminated composites under indentation loading using Acoustic Emission. *Compos. Part B Eng.* **2018**, *144*, 206–219. [CrossRef]
27. Sobhani, A.; Saeedifar, M.; Najafabadi, M.A.; Fotouhi, M.; Zarouchas, D. The study of buckling and post-buckling behavior of laminated composites consisting multiple delaminations using acoustic emission. *Thin-Walled Struct.* **2018**, *127*, 145–156. [CrossRef]
28. Šofer, M.; Šofer, P.; Pagáč, M.; Volodarskaja, A.; Babiuch, M.; Gruň, F. Acoustic Emission Signal Characterisation of Failure Mechanisms in CFRP Composites Using Dual-Sensor Approach and Spectral Clustering Technique. *Polymers* **2023**, *15*, 47. [CrossRef] [PubMed]
29. Fotouhi, M.; Pashmforoush, F.; Ahmadi, M.; Refahi Oskouei, A. Monitoring the initiation and growth of delamination in composite materials using acoustic emission under quasi-static three-point bending test. *J. Reinf. Plast. Compos.* **2011**, *30*, 1481–1493. [CrossRef]
30. Guo, Y.; Zhu, S.; Chen, Y.; Liu, D.; Li, D. Acoustic Emission-Based Study to Characterize the Crack Initiation Point of Wood Fibre/HDPE Composites. *Polymers* **2019**, *11*, 701. [CrossRef]
31. Panasiuk, K.; Dudzik, K. Determining the Stages of Deformation and Destruction of Composite Materials in a Static Tensile Test by Acoustic Emission. *Materials* **2022**, *15*, 313. [CrossRef]
32. Panasiuk, K.; Dudzik, K.; Hajdukiewicz, G. Acoustic Emission as a Method for Analyzing Changes and Detecting Damage in Composite Materials During Loading. *Arch. Acoust.* **2021**, *46*, 399–407. [CrossRef]
33. Panasiuk, K.; Kyzioł, L.; Dudzik, K. The use of acoustic emission signal (AE) in mechanical tests. *Przegląd Elektrotechniczny* **2019**, *11*, 8–11. [CrossRef]
34. Garbacz, G.; Kyzioł, L. Application of metric entropy to determine properties of structural materials. *Polím. Ciênc. Tecnol.* **2020**, *29*, e2019050. [CrossRef]
35. Garbacz, G.; Kyzioł, L. Application of metric entropy for results interpretation of composite materials mechanical tests. *Adv. Mater. Sci.* **2017**, *17*, 70–81. [CrossRef]
36. Kyzioł, L.; Hajdukiewicz, G. Application of the Kolmogorov-Sinai Entropy in Determining the Yield Point, as Exemplified by the EN AW-7020 Alloy. *J. KONBiN* **2019**, *49*, 241–269. [CrossRef]
37. *PN-EN ISO 527-4:2023-10*; Plastics—Determination of Tensile Properties—Part 4: Test Conditions for Isotropic and Orthotropic Fiber-Reinforced Plastic Composites. ISO: Geneva, Switzerland, 2021.

materials

MDPI

Article

Stability and Failure of Thin-Walled Composite Plate Elements with Asymmetric Configurations

Katarzyna Falkowicz

Faculty of Mechanical Engineering, Lublin University of Technology, Nadbystrzycka 38, 20-618 Lublin, Poland; k.falkowicz@pollub.pl; Tel.: +48-81-538-4204

Abstract: In the present study, the stability and failure phenomena of thin-walled constructions subjected to axial compression, featuring a central cut-out, and constructed from composite materials were explored. These constructions were fabricated from a carbon–epoxy composite using the autoclave method. The research encompassed experimental assessments on actual specimens alongside numerical analyses employing the finite element approach within the ABAQUS® software. The investigation spanned the entire load spectrum up to the point of structural failure, incorporating both practical trials and simulation analysis. During the practical assessments, the study monitored the post-buckling response and captured acoustic emissions to thoroughly evaluate the composite's failure mechanisms. Additionally, the ARAMIS system's non-invasive three-dimensional scanning was employed to assess deformations. Theoretical simulations utilized a step-by-step failure analysis, initiating with failure onset as per Hashin's theory and proceeding to failure progression based on an energy criterion. The simulation outcomes, particularly concerning the critical and post-critical phases, were juxtaposed with empirical data to identify the composite's vulnerability zones. The comparison underscored a significant concordance between the simulation predictions and the empirical findings.

Keywords: FEM analysis; plate elements; thin-walled structures; failure analysis; failure of composites; critical and post-critical states

Citation: Falkowicz, K. Stability and Failure of Thin-Walled Composite Plate Elements with Asymmetric Configurations. *Materials* **2024**, *17*, 1943. https://doi.org/10.3390/ma17091943

Academic Editors: Luís Miguel Pereira Durão and Nuno Calçada Loureiro

Received: 13 March 2024
Revised: 17 April 2024
Accepted: 19 April 2024
Published: 23 April 2024

1. Introduction

Thin-walled structures form a critical part of load-bearing frameworks utilized in modern engineering, including aerospace [1–3], automotive, and construction applications [4–6]. These structures are distinguished by their superior strength characteristics, which contribute to a high load-bearing capacity while maintaining a minimal self-weight. The increasing adoption of composite materials, susceptible to complex load conditions, necessitates in-depth investigations. Analytical methods and assessments of the stress levels in composite structures are continually being enhanced, particularly concerning their load-bearing capabilities. The nature of thin-walled load-bearing structures means that under certain load conditions, individual elements may risk stability loss within operationally permissible loads. Hence, beyond strength criteria, specific rigidity requirements are imposed to prevent early destruction due to stability loss in their elements, as discussed in various studies [7–11].

The buckling effects observed in thin-walled profiles are generally undesirable; numerous studies have demonstrated that stability loss in the form of elastic buckling, with a stable post-buckling equilibrium path, allows for the continued bearing of compressive loads in the post-buckling range [12–14].

In the critical phase, thin-walled structures made of composites often do not show any apparent signs of damage because the structure typically maintains its capability to support axial compressive forces even after surpassing the critical load. The functionality of these constructions is commonly linked with the likelihood of damage to the composite structure occurring predominantly in the extended post-critical operational phase [15,16].

The emergence of damage in this advanced post-critical stage can markedly diminish the mechanical integrity of the composite structure, thus posing a risk to its safe functionality even within the bounds of operationally acceptable loads. The multifaceted nature of damage and failure in thin-walled composite constructions is highlighted by a variety of phenomena that accompany irreversible structural damage, including the breakage of fibers or matrix and delamination [17–24].

The complexity of damage and destruction phenomena within composite material structures necessitates thorough analysis within ongoing research. Analyzing the full spectrum of loads initiates with a phase known as the damage initiation of the composite material, followed by a progression toward load-bearing capacity loss directly associated with damage evolution. The foundational theory for describing in detail the failure of thin-walled composite elements is known as the First Ply Failure theory (FPF) [25–29]. According to this theory, the composite is deemed to fail upon the damage state reaching the first layer of the laminate. This can be observed during experimental studies that utilize acoustic signal recording techniques (count numbers, hits numbers, signal amplitude, or energy) based on the acoustic emission method [30–34]. For numerical calculations, damage initiation assessment can be conducted using established damage initiation criteria, where Hashin's damage initiation criterion is of particular significance [35]. This criterion assesses the damage initiation state based on reaching the initial damage parameters of the composite material, including fiber damage due to tensile and compression forces and matrix damage due to tensile and compression forces. Moreover, utilizing Hashin's criterion [10,36–38] allows for further analysis within the subsequent damage evolution (with additional use of the energy criterion).

The preliminary concept posits that the damage can be conceptualized as the emergence of specific micro-fractures or the diminution of the effective cross-sectional area due to these micro-fractures, as introduced by Kachanov [39] in his delineation of destruction. This perspective envisages that the structure's load is conveyed solely through its intact cross-sectional area. The prevalent discourse in the literature focuses on Progressive Damage Analysis (PDA) [40–42], with the onset of damage being identified based on Hashin's criterion discussed earlier. PDA necessitates the specification of a suitable material model incorporating parameters for the initiation and progression of damage. The diminishment in stiffness following the onset of damage adheres to the framework put forth by Matzenmiller [43]. The gradual decrement in material stiffness is governed by variable parameters designed for the progression of damage. PDA identifies five key elements corresponding to the deterioration of material characteristics: damage under tensile and compressive forces in fibers, damage under tensile and compressive forces in the matrix, and damage from interlaminar shear [8,44].

This study provides a comprehensive experimental and numerical analysis of the buckling and subsequent post-buckling responses of compressed thin-walled composite plates, with a particular emphasis on asymmetric layouts. Additionally, this paper employs multidisciplinary research techniques that facilitate the examination of nonlinear stability problems in conjunction with the destruction phase of the composite material. This study meticulously documents the experimental process, which involves axial compression tests complemented by acoustic emission techniques to track damage evolution. Numerical simulations, employing finite element analysis (FEA) through ABAQUS® 2024, reinforce the experimental findings by providing detailed insights into the critical and post-critical states of the structures. This dual approach not only confirms the experimental observations but also highlights the significance of progressive failure analysis in understanding the material and structural integrity under compressive load.

A novelty in terms of research activity was the use of asymmetric systems and the mechanical couplings occurring in them in order to obtain an element that can work as an elastic element. The research discussed herein stemmed from the necessity to understand the impact of mechanical coupling in plate elements functioning as elastic components. Engineering machines and devices occasionally demand the incorporation of protective elements with minimal weight and precise operational traits. Furthermore, designing machines often involves the integration of elastic components within confined, rectangular spaces. In both scenarios, plates featuring cut-outs prove to be a viable solution. Previous research works on plate elements [11,45] mainly covered the critical and low-critical states. Therefore, the novelty of this work is also the description of the phenomenon of destruction using interdisciplinary research methods.

2. Study Subject and Material Properties

The focus of this research was on thin-walled plates made from carbon–epoxy composite material. The tested plates consisted of 12 layers in asymmetrical configuration arrangements relative to the laminate middle plane: $[\alpha/-\alpha_2/\alpha/0/\alpha/-\alpha/\alpha/-\alpha_2/\alpha/0]_T$, where α means the angle of fiber arrangement and equaled 30°, 45°, and 60°. Each layer's thickness was precisely 0.105 mm. Every tested plate element featured a central rectangular cut-out measuring 40 mm in width (b) and 100 mm in length (a). The plate geometry and composite configuration were informed by earlier studies [46–48]. Enhancements to the performance of slender plate elements and the assurance of their work within the elastic regime post-buckling were achieved through the introduction of a central cut-out and an asymmetrical layering technique. The determination of the layer arrangement stemmed from selecting suitable mechanical couplings from the B matrix, wherein the tested scenario involves strips and a core characterized by $A_S B_T D_S$ mechanical couplings. The methodology for choosing mechanical couplings has been elaborated further in several studies [46,49–51]. This method led to the creation of a novel elastic component concept. Such a strategy facilitated the bending of the plate's vertical segments in reverse directions during the compression tests [46]. The geometry and layup configurations of samples are presented in Figure 1. The tested samples were manufactured using the autoclave method under special and sterile conditions. This method facilitated the creation of a composite structure featuring a unique fiber content percentage within the range of 55–60%.

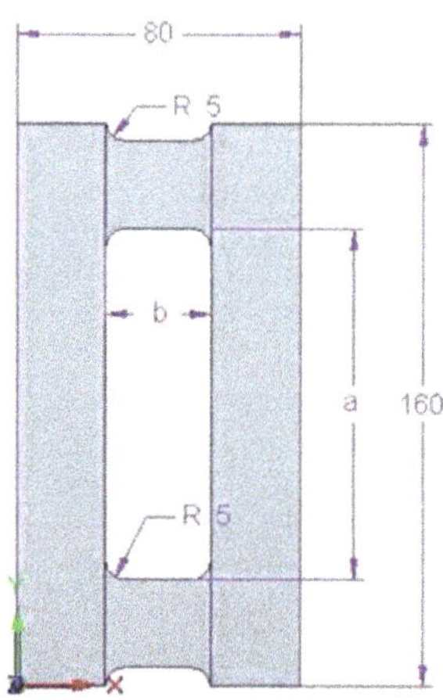

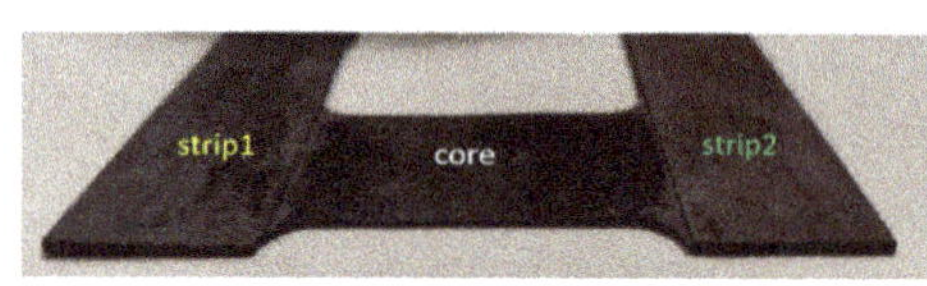

Name	Configuration
strip1	$[\alpha/-\alpha_2/\alpha/0/\boldsymbol{\alpha}/\boldsymbol{-\alpha}/\alpha/-\alpha_2/\alpha/0]_T$
strip2	$[0/\alpha/-\alpha_2/\alpha/\boldsymbol{\alpha}/\boldsymbol{-\alpha}/0/\alpha/-\alpha_2/\alpha]_T$
core	$[\boldsymbol{\alpha}/\boldsymbol{-\alpha}]$

Name	Final configuration
PN_α	$[\alpha/-\alpha_2/\alpha/0/\alpha/-\alpha/\alpha/-\alpha_2/\alpha/0]_T$

Figure 1. Geometry and layup configuration.

The material properties were experimentally established in accordance with ISO standards: the compression test followed PN-EN ISO 14126 (from 2002) [52], the static tensile test adhered to PN-EN ISO 527–5 (from 2010) [53], and the shear test complied with PN-EN ISO 14129 (from 2000) [54]. Details on the procedure for determining material properties are outlined in [55]. The material properties are presented in Table 1.

Table 1. Mechanical and strength properties of composite material.

Mechanical Properties				
Young's modulus E_1 [MPa]	Young's modulus E_2 [MPa]	Poisson's ratio [-]		Kirchhoff modulus G_{12} [MPa]
143,530	5826	0.36		3845
Strength Properties				
Tensile strength (0°) F_{T1} [MPa]	Compressive strength (0°) F_{C1} [MPa]	Tensile strength (90°) F_{T2} [MPa]	Compressive strength (90°) F_{C2} [MPa]	Shear strength F_{12} [MPa]
2221	641	49	114	83.5

3. Experimental Test

Thin-walled CFRP (Carbon Fiber-Reinforced Polymer) plates underwent compressive damage testing. The experiments were conducted using the Instron universal testing machine under ambient conditions at a temperature of 23 °C and a constant crosshead speed of 2 mm/min. The testing apparatus was specially outfitted with designed grips on both the upper and lower crossheads to ensure the samples received articulated support (refer to Figure 2). The grips had cylindrical inserts with notches where the plate could be mounted. The inserts were made of Teflon material, which ensured a low friction coefficient during the tests. This setup closely aligned with the freely supported boundary conditions assumed in the numerical analysis [56]. Testing on physical models of thin-walled elements, which were fabricated using the autoclave technique [57–59], facilitated the validation of the numerical model.

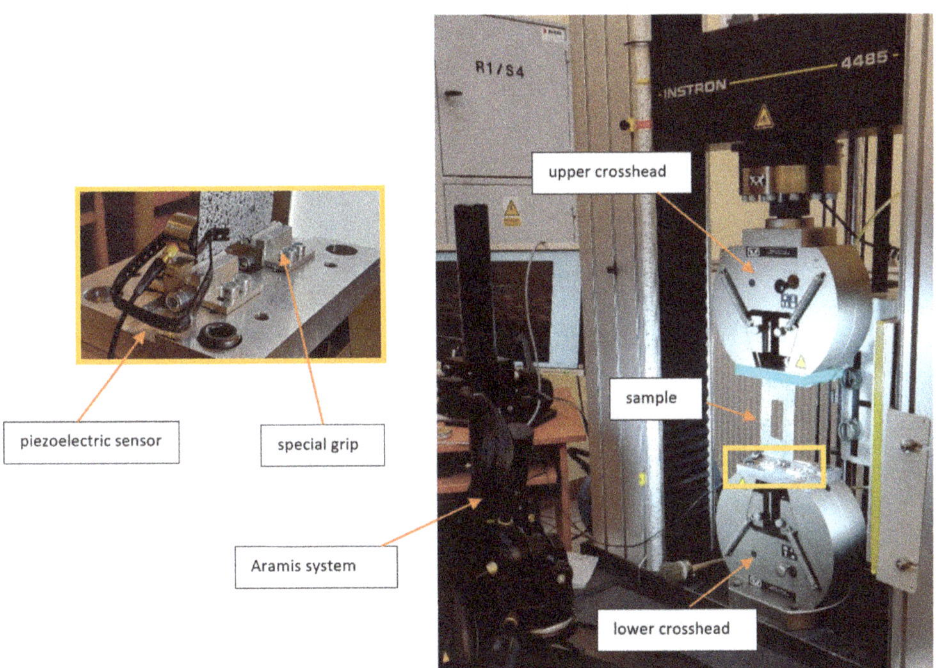

Figure 2. Experiment test stand for compression tests.

Experimental tests were conducted on composite samples, with the laminate ply configuration as previously described, at three distinct fiber orientations: 30°, 45°, and 60°. The tests measured several parameters: the compressive force applied, the shortening of the plate, the deflection in the direction perpendicular to the plate surface (notably at the mid-height of the plate's vertical strips, where deflection peaked), and the acoustic emission signals. For measuring deflection, ARAMIS by GOM was utilized [60,61], offering comprehensive visualization of displacements and strains. Acoustic emission signals [30] were captured using an AMSY-5 system equipped with a Fujicera 1045S piezoelectric sensor and an AEP-4 signal amplifier, allowing for the documentation of the initial damage indicators in terms of energy, amplitude, counts, and sum of counts. The experimental outcomes were then juxtaposed with the plate shortening predictions derived from numerical simulations (FEM). The author has elaborated on the experimental procedures in earlier publications [62,63].

In the course of the experimental tests, two critical forces were identified: the force initiating damage in the first composite layer (P_d), and the failure force (P_f), which corresponds to the point at which the structure loses its load-bearing capacity.

4. Numerical Analysis

The numerical analyses conducted in Abaqus software 2024 are grounded in the Finite Element Method (FEM), renowned for its extensive applicability [2,64–70]. This analysis of compressed composite plates unfolded in two distinct phases. Initially, a linear eigenproblem was addressed based on the minimum potential energy principle [71] Subsequently, the second phase of numerical analysis involved conducting nonlinear stability assessments employing the Newton–Raphson incremental–iterative method [72–74]. These assessments accounted for geometric imperfections aligning with the flexural–torsional buckling mode. The research utilized six-node shell elements, each endowed with six degrees of freedom. The geometric imperfection used in the analysis equaled 0.1 of the plate thickness. Abaqus software's section configuration capabilities facilitated the distinct definition of each laminate ply, including material properties, thickness, and the principal axes of the orthotropic material. Material parameters are detailed in Table 1, where the properties of the composite material were characterized by an orthotropic material model under plane stress conditions.

The plate elements' geometry was meticulously crafted to mirror the dimensions of the actual specimens, as depicted in Figure 1. To accurately replicate the experimental test conditions, two rigid plates were simulated to serve as the top and bottom edge supports for the test plate. These support plates were modeled using rigid shell elements of the R3D4 type, each endowed with three translational degrees of freedom at its four computational nodes within each finite element. Reference points were established in conjunction with these non-deformable plates, where the boundary conditions were specified (see Figure 3). The constructed numerical model comprised 4244 finite elements, ensuring a detailed representation of the physical testing setup. The element size of 1.5 mm was used in subsequent analysis. Mesh convergence analysis was performed in previous papers [75].

Numerical analyses were carried out across the entire spectrum of loads, culminating in total failure, employing Progressive Failure Analysis (PFA) grounded in the Hashin failure criterion and an energy criterion pertinent to damage evolution [8,35,44,76]. This criterion enables the evaluation of damage levels within the composite material attributable to fiber tension (HSNFTCRT), fiber compression (HSNFCCRT), matrix tension (HSNMTCRT), and matrix compression (HSNMCCRT). When any of these damage initiation criteria meet or exceed a value of 1 for the failure initiation parameter, it indicates the potential for further damage progression within a specific material component via PFA, while the rest of the components remain intact.

In line with Hashin's theory, the activation of a damage initiation criterion for any component of the composite triggers a gradual reduction in that component's stiffness, leading eventually to a total loss of material stiffness. The damage evolution model incorporates five independent failure parameters for the composite material, each corresponding to failure modes such as fiber compression (DAMAGEFC), fiber tension (DAMAGEFT), matrix compression (DAMAGEMC), matrix tension (DAMAGEMT), and interlaminar shear (DAMAGESHR).

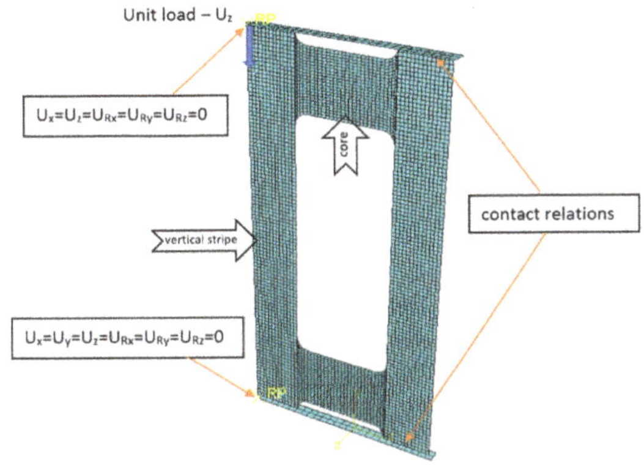

Figure 3. FE model with boundary conditions.

Further details on the methodologies applied can be found in the referenced literature [62,63,75,77,78].

5. Results and Discussion

The investigation into the buckling behaviors of compressed composite plates with central cut-outs, through both experimental and numerical approaches, revealed a qualitative concurrence. Moreover, the analytical calculations were used to indicate A, B, and D matrix values to confirm a coupling effect [79–81]. Experimental analyses verified that the primary mode of buckling was indeed flexural–torsional, demonstrating stability within the post-buckling range. Moreover, the analysis of the critical state, via numerical calculations, facilitated the identification of various buckling forms and the corresponding critical loads for each specific layer configuration within the composite material, as illustrated in Figure 4. Such observations affirm the efficacy of the chosen asymmetric fiber arrangement and the mechanical couplings. Equations (1)–(3) present the results of analytical calculations of constitutive relations for considered configurations. The calculations were performed using the Matlab program.

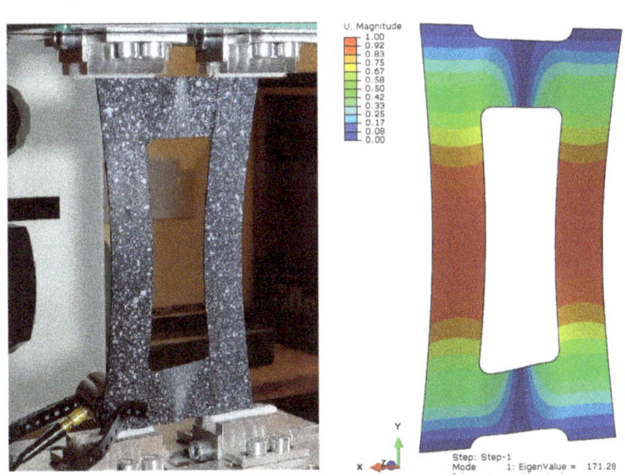

(a)

Figure 4. *Cont.*

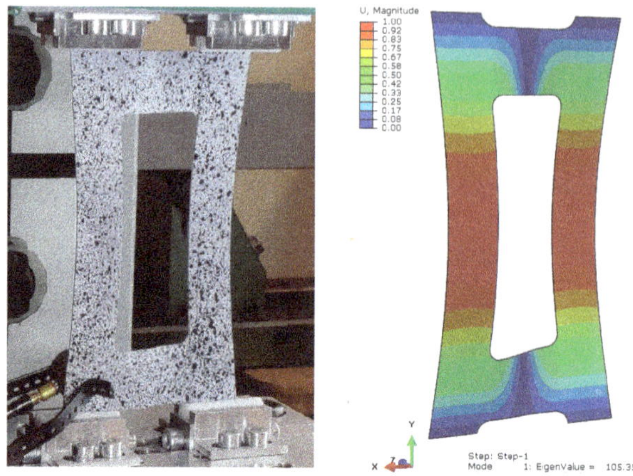

(**b**)

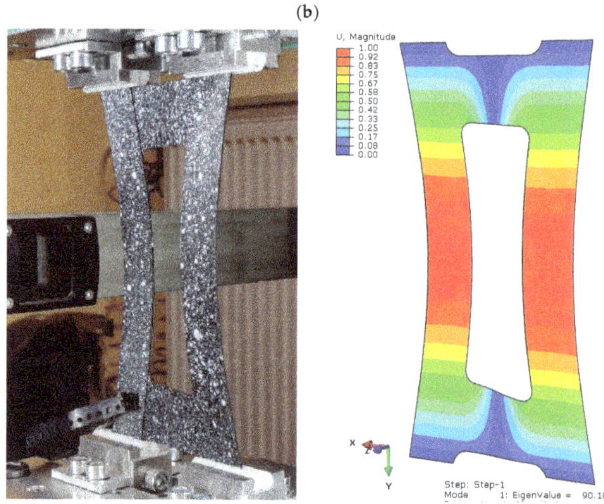

(**c**)

Figure 4. Experimental and numerical buckling form: PN_30 (**a**); PN_45 (**b**); PN_60 (**c**).

The ABD matrix for the PN_30, PN_45, and PN_60 configuration is given in Equations (1)–(3), respectively, and represents E-B-S-B-E-T-S-T-T-B coupling.

$$\left\{\begin{array}{c}\left\{\begin{array}{c}N_x\\N_y\\N_{xy}\end{array}\right\}\\ \cdots\\ \left\{\begin{array}{c}M_x\\M_y\\M_{xy}\end{array}\right\}\end{array}\right\}=\left[\begin{array}{ccccccc}119763.91 & 28358.47 & 0 & \vdots & 2605.69 & -1079.48 & -486.24\\ 28358.47 & 18016.22 & 0 & \vdots & -1079.48 & -446.74 & -174.63\\ 0 & 0 & 30546.52 & \vdots & -486.24 & -174.63 & -1079.48\\ \cdots & \cdots & \cdots & \cdots & \cdots & \cdots & \cdots\\ & & & \vdots & 16437.56 & 3506.25 & 408.45\\ & symmetry & & \vdots & 3506.25 & 2281.91 & 146.69\\ & & & \vdots & 408.45 & 146.69 & 3795.72\end{array}\right]=\left\{\begin{array}{c}\left\{\begin{array}{c}\varepsilon_x\\ \varepsilon_y\\ \gamma_{xy}\end{array}\right\}\\ \cdots\\ \left\{\begin{array}{c}\mathcal{K}_x\\ \mathcal{K}_y\\ \mathcal{K}_{xy}\end{array}\right\}\end{array}\right\} \quad (1)$$

$$\left\{\begin{Bmatrix} N_x \\ N_y \\ N_{xy} \\ \cdots \end{Bmatrix} \atop \begin{Bmatrix} M_x \\ M_y \\ M_{xy} \end{Bmatrix}\right\} = \begin{bmatrix} 74858.17 & 36925.75 & 0 & \vdots & 4491.73 & -1439.3 & -381.55 \\ 36925.75 & 45787.4 & 0 & \vdots & -1439.3 & -1613.13 & -381.55 \\ 0 & 0 & 39113.8 & \vdots & -381.55 & -381.55 & -1439.3 \\ \cdots & \cdots & \cdots & \cdots & \cdots & \cdots & \cdots \\ & & & \vdots & 10925.61 & 4557.84 & 320.51 \\ & symmetry & & \vdots & 4557.84 & 5690.69 & 320.51 \\ & & & \vdots & 320.51 & 320.51 & 4847.31 \end{bmatrix} = \left\{\begin{Bmatrix} \varepsilon_x \\ \varepsilon_y \\ \gamma_{xy} \\ \cdots \end{Bmatrix} \atop \begin{Bmatrix} \mathcal{K}_x \\ \mathcal{K}_y \\ \mathcal{K}_{xy} \end{Bmatrix}\right\} \quad (2)$$

$$\left\{\begin{Bmatrix} N_x \\ N_y \\ N_{xy} \\ \cdots \end{Bmatrix} \atop \begin{Bmatrix} M_x \\ M_y \\ M_{xy} \end{Bmatrix}\right\} = \begin{bmatrix} 47086.99 & 28358.47 & 0 & \vdots & 5658.12 & -1079.48 & -174.63 \\ 28358.47 & 90693.14 & 0 & \vdots & -1079.48 & -3499.17 & -486.24 \\ 0 & 0 & 30546.52 & \vdots & -174.63 & -486.24 & -1079.48 \\ \cdots & \cdots & \cdots & \cdots & \cdots & \cdots & \cdots \\ & & & \vdots & 7516.83 & 3506.25 & 146.69 \\ & symmetry & & \vdots & 3506.25 & 11202.64 & 408.45 \\ & & & \vdots & 146.69 & 408.45 & 3795.72 \end{bmatrix} = \left\{\begin{Bmatrix} \varepsilon_x \\ \varepsilon_y \\ \gamma_{xy} \\ \cdots \end{Bmatrix} \atop \begin{Bmatrix} \mathcal{K}_x \\ \mathcal{K}_y \\ \mathcal{K}_{xy} \end{Bmatrix}\right\} \quad (3)$$

The considered configurations indicate developed *Bending–Twisting* behaviour (B_{16}, $B_{26} \neq 0$), and the addition of *Bending–Twisting* coupling stiffnesses D_{16} and D_{26} in laminate causes *Twisting* because of *Bending*. Moreover, Figure 5 presents the relations between compression force and the rotation value, measurements for the plate strip on one side and on the other side.

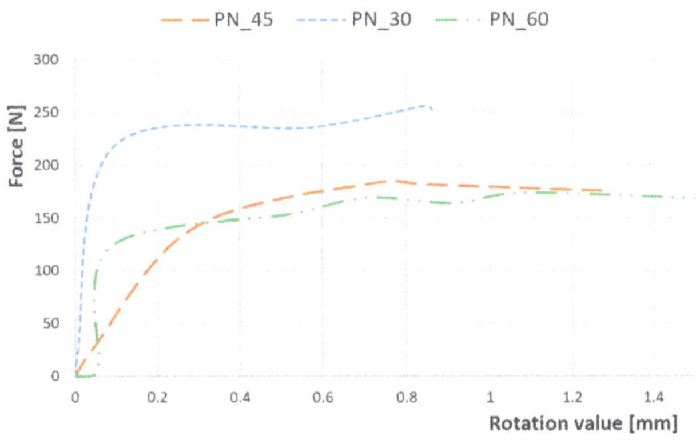

Figure 5. Results for compression load vs. rotation value for three different angles.

The rotational value shows a notable increase, particularly when the load approaches the buckling threshold. Additionally, there is a discernible rise in the twisting response of the specimen upon the introduction of laminate coupling effects. The results obtained provide clear evidence of coupling interactions occurring, which contribute to the amplified twisting response.

The examination of the structure's response to compression enabled the identification of post-critical working paths (compressive load versus displacement/shortening). The analysis revealed that these post-critical equilibrium paths exhibit a stable nature, indicating that an increase in the shortening of the thin-walled structure is paralleled by a rise in compressive load, even in the face of stability loss—as illustrated in Figure 6. Within the post-critical domain, continuous loading of the structure led to a phase termed damage initiation. During the experimental phase, parameters of the acoustic emission signal, including the number of counts, energy, and amplitudes, were meticulously recorded. These data subsequently facilitated the evaluation of the onset of damage. The results obtained are shown in Figures 7–9. It is noteworthy that experimentally, only sample PN_30 succeeded in bringing the sample to complete failure.

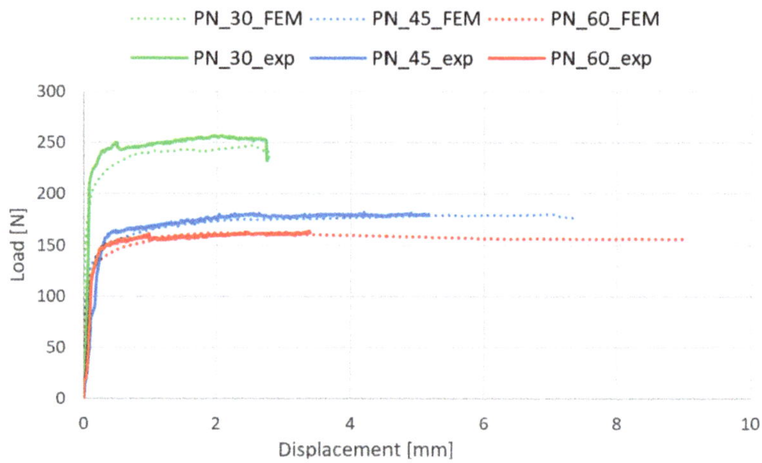

Figure 6. Comparison of experimental and numerical post-buckling equilibrium paths.

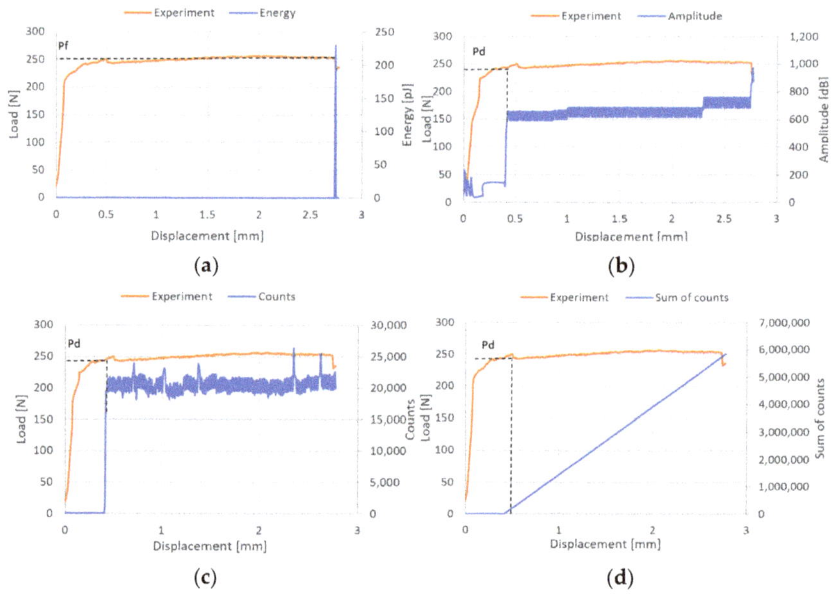

Figure 7. Post-critical equilibrium paths with acoustic emission signals for PN_30: (**a**) energy; (**b**) amplitude; (**c**) counts; (**d**) sum of counts.

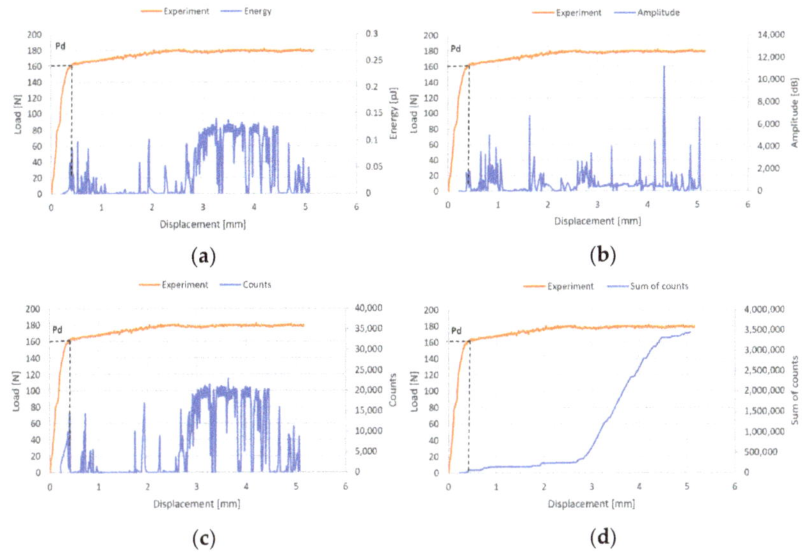

Figure 8. Post-critical equilibrium paths with acoustic emission signals for PN_45: (**a**) energy; (**b**) amplitude; (**c**) counts; (**d**) sum of counts.

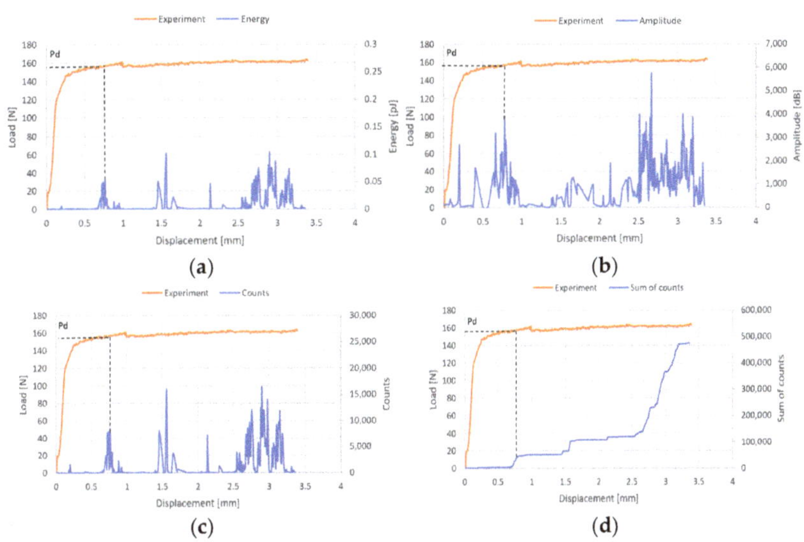

Figure 9. Post-critical equilibrium paths with acoustic emission signals for PN_60: (**a**) energy; (**b**) amplitude; (**c**) counts; (**d**) sum of counts.

Based on the results obtained, it turned out that the sum of counts was the clearest acoustic emission signal in assessing the initiation of damage. On the other hand, for the loss of load-bearing capacity of the structure, thanks to the occurrence of a characteristic "peak" of energy, the energy parameter can also be recognized. However, it could only be recorded for specimen PN_30, as the other specimens failed to initiate damage. Upon examining the amplitude charts, it becomes evident that the amplitude of the acoustic emission signals exhibits a trend of increment in their recorded values as the duration of the measurement extends. This observation signifies that, with the progression of the testing period, the acoustic emissions, as quantified by their amplitudes, intensify. Such an escalation is typically indicative of an accumulation of damage within the material or structure under evaluation. By capturing these signals, the acoustic emission technique serves as a sensitive method to detect early and developing damages, like micro-cracking or the fracturing of fibers, within composite materials subjected to load. The increasing amplitudes imply that the material encounters more significant stress or damage events as the testing proceeds, offering valuable insights into the behavior of the material under sustained load conditions. It is worth noting that for samples PN_45 and PN_60, based on the results of a single signal, it would be difficult to unambiguously determine the value of the load initiating composite failure (P_d) compared to sample PN_30. It can be observed that P_d damage initiation for all samples occurred due to matrix tension—HSNMTCRT (Figure 10). Figure 10 presents the maps showing the level of damage evolution after the fulfillment of at least two components (PN_30:DAMAGEMT, DAMAGESHR; PN_45:DAMAGEMT, DAMAGEMC, DAMAGESHR; PN_60:DAMAGEMT, DAMAGEMC, DAMAGESHR, DAMAGEFC). The presented results confirm the influence of the fiber arrangement on both the value of the critical force and the initiation and failure evolution. Additionally, it is related to the influence of mechanical couplings in the chosen configuration. However, this work mainly focused on the stability and failure of thin-walled composite plate elements as well as the angle of the fiber arrangement.

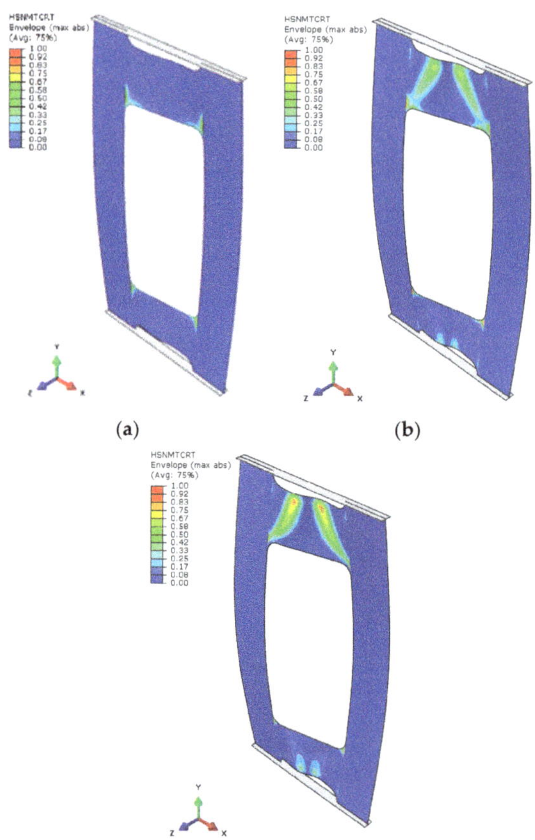

Figure 10. Damage initiation of plate (numerical analysis): PN_30 (**a**); PN_45 (**b**); PN_60 (**c**).

The area of damage observed in the experimental research aligns with the area predicted by the numerical simulations (Figure 11). In every scenario, the zones near the cut-out corners, as well as the core area, emerged as the most prone to damage, with destruction primarily initiated at the corners. A pattern was discernible, indicating that an increase in the fiber alignment angle tends to decrease the plate's rigidity, leading to stress concentration in the core. As previously noted, failure was only achieved in the case of the PN_30 plate. According to Figure 12, the material failure that compromised the plate's load-bearing capacity was located near the upper left corner of the cut-out and in the section of the core linking to the plate's vertical strip.

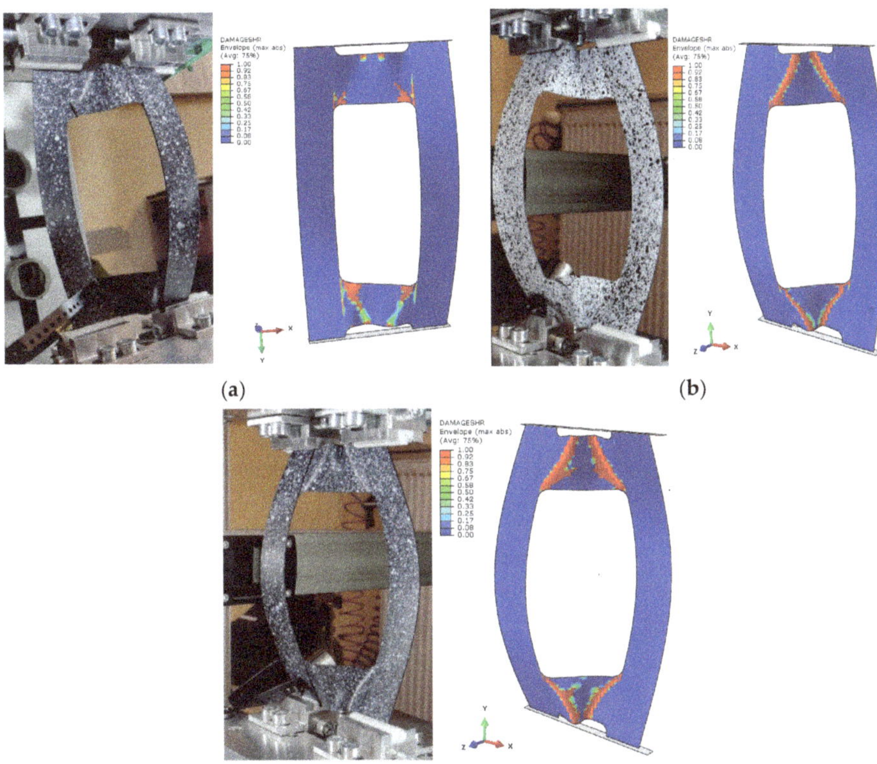

(**a**) (**b**) (**c**)

Figure 11. Comparison of the damage evolution results from experimental test and FEM analysis: PN_30 (**a**); PN_45 (**b**); PN_60 (**c**).

To provide a quantitative summary of the test outcomes across the three distinct fiber orientation configurations, the collected results have been compiled and are showcased in Table 2. In addition, the last columns include the value of the relative error with respect to the experimental values.

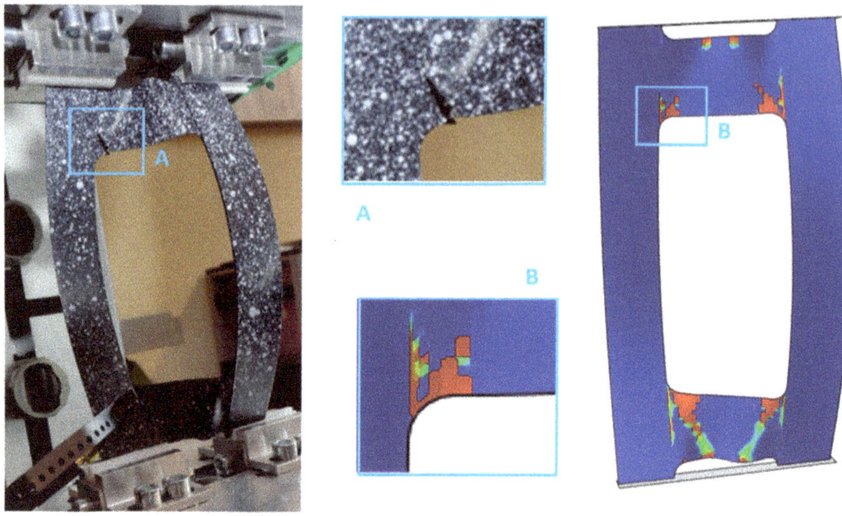

Figure 12. Graphical presentation of the form of failure of the plate in PN_30 configuration.

Table 2. Limit load results.

	Experimental Test [N]			FEM Analysis [N]			Error [%]		
	PN_30	PN_45	PN_60	PN_30	PN_45	PN_60	PN_30	PN_45	PN_60
Damage initiation	242	160.31	156.87	228.31	161.09	160.09	5.65	0.49	2.05
Failure	251.87	-	-	239.33	178.34	-	4.98	-	-

The limit state analysis showed that the relative error between experimental and numerical tests for the load initiating damage of the first composite layer was 5.65% for PN_30, 0.49% for PN_45, and 2.05% for PN_60. On the other hand, in the case of the load initiating delamination and the load causing a loss of load capacity for PN_30, it was 4.98%.

The results obtained allow for the evaluation of the accuracy in the development of the numerical model, which was corroborated by the outcomes of the experimental tests.

6. Conclusions

This paper presents the results of comprehensive experimental and numerical investigations into the stability and failure of thin-walled composite plates subjected to axial compression, focusing on asymmetric layer configurations. This research utilized carbon–epoxy composite materials crafted through the autoclave technique, meticulously exploring the entire load spectrum up to structural failure. By integrating experimental tests with numerical simulations via the finite element method in ABAQUS®, this study illuminated the complex failure mechanisms of the composites. Particularly, the research leveraged acoustic emission monitoring and the ARAMIS non-invasive scanning system for deformation assessment, alongside a progressive failure analysis based on Hashin's criterion and an energy criterion for detailed failure evolution mapping.

Key findings from this investigation include the validation of flexural–torsional buckling as the primary mode of failure, which was consistently observed in both experimental and numerical analyses. This buckling mode's identification underscores the effectiveness of the selected asymmetric fiber layout and mechanical couplings in ensuring the structure's stability beyond the critical load. Additionally, this study identified the critical and failure loads for various layer configurations, enhancing the understanding of how composite layer arrangements influence structural integrity.

A notable innovation in this research was the application of asymmetric layouts and the exploration of mechanical couplings within these structures to create elements capable of functioning as elastic components. This approach, which deviates from traditional studies focusing primarily on critical and low-critical states, offers a novel perspective on the destruction phenomenon using interdisciplinary methods.

In summary, this work contributes significantly to the field of composite structure analysis by providing a detailed examination of the buckling behavior and failure mechanisms of thin-walled composite plates with asymmetric configurations. The synergy between experimental insights and numerical predictions offers a robust framework for predicting composite structures' behavior under axial compression, paving the way for designing more resilient composite materials and structures in the future. This research was made possible by funding from the National Science Centre Poland, underlining its scientific significance and its impact on advancing composite materials research.

Funding: This research was conducted under project No. 2022/47/B/ST8/00600, financed by the National Science Centre, Poland. This research was funded in whole or in part by the National Science Centre, Poland [2022/47/B/ST8/00600]. For the purpose of Open Access, the author has applied a CC-BY public copyright license to any Author Accepted Manuscript (AAM) version arising from this submission.

Institutional Review Board Statement: Not applicable.

Informed Consent Statement: Not applicable.

Data Availability Statement: Data are contained within the article.

Conflicts of Interest: The author declares no conflicts of interest.

References

1. Kopecki, T.; Lis, T.; Mazurek, P. Post-Critical Deformation of Thin-Walled Load-Bearing Aircraft Structure Representing Fragment of the One-Way Torsion Box. *Adv. Sci. Technol. Res. J.* **2018**, *12*, 203–209. [CrossRef]
2. Święch, Ł. Finite Element Analysis of Stress Distribution in the Node Region of Isogrid Thin-Walled Panels. In *Modelling in Engineering 2020: Applied Mechanics*; Mężyk, A., Kciuk, S., Szewczyk, R., Duda, S., Eds.; Advances in Intelligent Systems and Computing; Springer International Publishing: Cham, Switzerland, 2021; Volume 1336, pp. 279–288. ISBN 978-3-030-68454-9.
3. Guo, S.; Cheng, W.; Cui, D. Aeroelastic Tailoring of Composite Wing Structures by Laminate Layup Optimization. *AIAA J.* **2006**, *44*, 3146–3150. [CrossRef]
4. Chróścielewski, J.; Miśkiewicz, M.; Pyrzowski, Ł.; Rucka, M.; Sobczyk, B.; Wilde, K. Modal Properties Identification of a Novel Sandwich Footbridge—Comparison of Measured Dynamic Response and FEA. *Compos. Part B Eng.* **2018**, *151*, 245–255. [CrossRef]
5. Chróścielewski, J.; Miśkiewicz, M.; Pyrzowski, Ł.; Sobczyk, B.; Wilde, K. A Novel Sandwich Footbridge—Practical Application of Laminated Composites in Bridge Design and in Situ Measurements of Static Response. *Compos. Part B Eng.* **2017**, *126*, 153–161. [CrossRef]
6. Schwab, M.; Pettermann, H.E. Modelling and Simulation of Damage and Failure in Large Composite Components Subjected to Impact Loads. *Compos. Struct.* **2016**, *158*, 208–216. [CrossRef]
7. Singer, J.; Arbocz, J.; Weller, T. *Buckling Experiments: Experimental Methods in Buckling of Thin-Walled Structures*; John Wiley & Sons, Inc.: New York, NY, USA, 2000; Volume 1, ISBN 978-0-470-17298-8.
8. Lapczyk, I.; Hurtado, J.A. Progressive Damage Modeling in Fiber-Reinforced Materials. *Compos. Part A Appl. Sci. Manuf.* **2007**, *38*, 2333–2341. [CrossRef]
9. Kopecki, T.; Mazurek, P.; Lis, T.; Chodorowska, D. Post-Buckling Deformation States of Semi-Monocoque Cylindrical Structures with Large Cut-Outs under Operating Load Conditions. Numerical Analysis and Experimental Tests. *EiN* **2016**, *18*, 16–24. [CrossRef]
10. Banat, D.; Mania, R.J.; Degenhardt, R. Stress State Failure Analysis of Thin-Walled GLARE Composite Members Subjected to Axial Loading in the Post-Buckling Range. *Compos. Struct.* **2022**, *289*, 115468. [CrossRef]
11. Wysmulski, P.; Falkowicz, K.; Filipek, P. Buckling State Analysis of Compressed Composite Plates with Cut-Out. *Compos. Struct.* **2021**, *274*, 114345. [CrossRef]
12. Wysmulski, P. The Effect of Load Eccentricity on the Compressed CFRP Z-Shaped Columns in the Weak Post-Critical State. *Compos. Struct.* **2022**, *301*, 116184. [CrossRef]
13. Falkowicz, K. Linear Analysis of Thin-Walled Composite Profiles Weakened by Holes. *Acta Mech. Autom.* **2023**, *17*, 550–557. [CrossRef]
14. Wysmulski, P.; Debski, H.; Falkowicz, K. Sensitivity of Compressed Composite Channel Columns to Eccentric Loading. *Materials* **2022**, *15*, 6938. [CrossRef] [PubMed]

15. van der Heijden, A.M.A.; Koiter, W.T. (Eds.) *W.T. Koiter's Elastic Stability of Solids and Structures*; Cambridge University Press: Cambridge, UK, 2009; ISBN 978-0-521-51528-3.
16. Banat, D.; Mania, R.J. Progressive Failure Analysis of Thin-Walled Fibre Metal Laminate Columns Subjected to Axial Compression. *Thin-Walled Struct.* **2018**, *122*, 52–63. [CrossRef]
17. Banat, D.; Mania, R.J. Damage Analysis of Thin-Walled GLARE Members under Axial Compression—Numerical and Experiment Investigations. *Compos. Struct.* **2020**, *241*, 112102. [CrossRef]
18. Banat, D.; Mania, R. Failure Analysis of Thin-Walled GLARE Members during Buckling and Post-Buckling Response. *AIP Conf. Proc.* **2019**, *2060*, 020001. [CrossRef]
19. Czajka, B.; Różyło, P.; Dębski, H. Stability and Failure of Thin-Walled Composite Structures with A Square Cross-Section. *ACS* **2022**, *18*, 43–55. [CrossRef]
20. Wysmulski, P. Numerical and Experimental Study of Crack Propagation in the Tensile Composite Plate with the Open Hole. *Adv. Sci. Technol. Res. J.* **2023**, *17*, 249–261. [CrossRef] [PubMed]
21. Wysmulski, P. Failure Mechanism of Tensile CFRP Composite Plates with Variable Hole Diameter. *Materials* **2023**, *16*, 4714. [CrossRef] [PubMed]
22. Wang, C.; Vassilopoulos, A.P.; Keller, T. Numerical Investigation of Two-Dimensional Mode-II Delamination in Composite Laminates. *Compos. Part A Appl. Sci. Manuf.* **2024**, *179*, 108012. [CrossRef]
23. Mrunmayee, S.; Rajagopal, A.; Rakesh, K.; Basant, K.; Reddy, J.N. Phase Field Approach to Predict Mixed-Mode Delamination and Delamination Migration in Composites. *Compos. Struct.* **2024**, *337*, 118074. [CrossRef]
24. Syed Abdullah, S.I.B.; Bokti, S.K.; Wong, K.J.; Johar, M.; Chong, W.W.F.; Dong, Y. Mode II and Mode III Delamination of Carbon Fiber/Epoxy Composite Laminates Subjected to a Four-Point Bending Mechanism. *Compos. Part B Eng.* **2024**, *270*, 111110. [CrossRef]
25. Reddy, J.N.; Pandey, A.K. A First-Ply Failure Analysis of Composite Laminates. *Comput. Struct.* **1987**, *25*, 371–393. [CrossRef]
26. Kam, T.Y.; Jan, T.B. First-Ply Failure Analysis of Laminated Composite Plates Based on the Layerwise Linear Displacement Theory. *Compos. Struct.* **1995**, *32*, 583–591. [CrossRef]
27. Jalalvand, M.; Fotouhi, M.; Wisnom, M.R. Effect of Shear Stresses on Fibre Direction Tensile Failure Using a New Simple and Reliable Test Method with Thin Plies. *Compos. Part A Appl. Sci. Manuf.* **2024**, *182*, 108155. [CrossRef]
28. Kovács, L.; Bugár-Mészáros, M.; Romhány, G. Holistic Estimation and Sensitivity-Based Experiment Design Method of Composite Laminate First-Ply Failure Models Using Statistical Approaches. *Eng. Fail. Anal.* **2024**, *156*, 107834. [CrossRef]
29. Chatterjee, D.; Ghosh, A.; Chakravorty, D. Nonlinear First Ply Failure Study of Laminated Composite Skew Plates. *Mater. Today Proc.* **2021**, *45*, 4925–4930. [CrossRef]
30. Kubiak, T.; Samborski, S.; Teter, A. Experimental Investigation of Failure Process in Compressed Channel-Section GFRP Laminate Columns Assisted with the Acoustic Emission Method. *Compos. Struct.* **2015**, *133*, 921–929. [CrossRef]
31. Almeida, R.S.M.; Magalhães, M.D.; Karim, M.N.; Tushtev, K.; Rezwan, K. Identifying Damage Mechanisms of Composites by Acoustic Emission and Supervised Machine Learning. *Mater. Des.* **2023**, *227*, 111745. [CrossRef]
32. Zhou, W.; Zhao, W.; Zhang, Y.; Ding, Z. Cluster Analysis of Acoustic Emission Signals and Deformation Measurement for Delaminated Glass Fiber Epoxy Composites. *Compos. Struct.* **2018**, *195*, 349–358. [CrossRef]
33. Pourdadash Fardnam, A.; Maleki, A.; Ahmadi Najafabadi, M. Investigation of Buckling of Laminated Composites Repaired by Resin Injection Using Acoustic Emission and Cohesive Zone Simulation Method. *Compos. Struct.* **2022**, *298*, 116008. [CrossRef]
34. Kempf, M.; Skrabala, O.; Altstädt, V. Acoustic Emission Analysis for Characterisation of Damage Mechanisms in Fibre Reinforced Thermosetting Polyurethane and Epoxy. *Compos. Part B Eng.* **2014**, *56*, 477–483. [CrossRef]
35. Hashin, Z.; Rotem, A. A Fatigue Failure Criterion for Fiber Reinforced Materials. *J. Compos. Mater.* **1973**, *7*, 448–464. [CrossRef]
36. Duarte, A.P.C.; Díaz Sáez, A.; Silvestre, N. Comparative Study between XFEM and Hashin Damage Criterion Applied to Failure of Composites. *Thin-Walled Struct.* **2017**, *115*, 277–288. [CrossRef]
37. Gu, J.; Chen, P. Some Modifications of Hashin's Failure Criteria for Unidirectional Composite Materials. *Compos. Struct.* **2017**, *182*, 143–152. [CrossRef]
38. Al-Hassani, S.T.S.; Darvizeh, M.; Haftchenari, H. An Analytical Study of Buckling of Composite Tubes with Various Boundary Conditions. *Compos. Struct.* **1997**, *39*, 157–164. [CrossRef]
39. Kachanov, L.M. Rupture Time Under Creep Conditions. *Int. J. Fract.* **1999**, *97*, 11–18. [CrossRef]
40. Bai, N.; Li, H.; Lan, C.; Spencer, B.F. Influencing Factors and Sensitivity Analysis for the Fatigue of FRP Wire Based on the Progressive Fatigue Damage Model. *Compos. Struct.* **2024**, *334*, 117982. [CrossRef]
41. Trombini, M.; Enea, M.; Arruda, M.R.T.; Pagani, A.; Petrolo, M.; Carrera, E. 1D Higher-Order Theories for Quasi-Static Progressive Failure Analysis of Composites Based on a Full 3D Hashin Orthotropic Damage Model. *Compos. Part B Eng.* **2024**, *270*, 111120. [CrossRef]
42. Liu, Z.; Yan, L.; Wu, Z.; Zhou, J.; Wei, H.; Zhang, S.; Ren, X. Progressive Damage Analysis and Experiments of Open-Hole Composite Laminates Subjected to Compression Loads. *Eng. Fail. Anal.* **2023**, *151*, 107379. [CrossRef]
43. Matzenmiller, A.; Lubliner, J.; Taylor, R.L. A Constitutive Model for Anisotropic Damage in Fiber-Composites. *Mech. Mater.* **1995**, *20*, 125–152. [CrossRef]
44. Li, W.; Cai, H.; Li, C.; Wang, K.; Fang, L. Progressive Failure of Laminated Composites with a Hole under Compressive Loading Based on Micro-Mechanics. *Adv. Compos. Mater.* **2014**, *23*, 477–490. [CrossRef]

45. Falkowicz, K.; Debski, H.; Teter, A. Design Solutions for Improving the Lowest Buckling Loads of a Thin Laminate Plate with Notch. *AIP Conf. Proc.* **2018**, *1922*, 080004. [CrossRef]
46. Falkowicz, K.; Debski, H.; Wysmulski, P. Effect of Extension-Twisting and Extension-Bending Coupling on a Compressed Plate with a Cut-Out. *Compos. Struct.* **2020**, *238*, 111941. [CrossRef]
47. Falkowicz, K.; Samborski, S.; Valvo, P.S. Effects of Elastic Couplings in a Compressed Plate Element with Cut-Out. *Materials* **2022**, *15*, 7752. [CrossRef]
48. Falkowicz, K.; Debski, H. The Work of a Compressed, Composite Plate in Asymmetrical Arrangement of Layers. *AIP Conf. Proc.* **2019**, *2078*, 020005. [CrossRef]
49. Falkowicz, K.; Dębski, H.; Wysmulski, P.; Różyło, P. The Behaviour of Compressed Plate with a Central Cut-out, Made of Composite in an Asymmetrical Arrangement of Layers. *Compos. Struct.* **2019**, *214*, 406–413. [CrossRef]
50. York, C.B. Unified Approach to the Characterization of Coupled Composite Laminates: Benchmark Configurations and Special Cases. *J. Aerosp. Eng.* **2010**, *23*, 219–242. [CrossRef]
51. York, C.B. Tapered Hygro-Thermally Curvature-Stable Laminates with Non-Standard Ply Orientations. *Compos. Part A Appl. Sci. Manuf.* **2013**, *44*, 140–148. [CrossRef]
52. *PN-EN ISO 14126*; Fiber-Reinforced Plastic Composites—Determination of Compression Properties Parallel to the Lamination Plane. International Organization for Standardization: Geneva, Switzerland, 2002.
53. *PN-EN ISO 527–5*; Plastics—Determination of Tensile Properties—Part 5: Test Conditions for Unidirectional Fiber-Reinforced Plastic Composites. International Organization for Standardization: Geneva, Switzerland, 2010.
54. *PN-EN ISO 14129*; Determination of Shear Stress and Corresponding Strain, Shear Modulus and Tensile Strength at an Angle of ±45° of Fiber-Reinforced Plastic Composites. International Organization for Standardization: Geneva, Switzerland, 2000.
55. Różyło, P.; Smagowski, W.; Paśnik, J. Experimental Research in the Aspect of Determining the Mechanical and Strength Properties of the Composite Material Made of Carbon-Epoxy Composite. *Adv. Sci. Technol. Res. J.* **2023**, *17*, 232–246. [CrossRef]
56. Falkowicz, K.; Wysmulski, P.; Debski, H. Buckling Analysis of Laminated Plates with Asymmetric Layup by Approximation Method. *Materials* **2023**, *16*, 4948. [CrossRef]
57. Wysmulski, P. Analysis of the Effect of an Open Hole on the Buckling of a Compressed Composite Plate. *Materials* **2024**, *17*, 1081. [CrossRef] [PubMed]
58. Yue, B.; Xu, Y.; Teng, X.; Zhou, Y.; He, L.; Ma, S.; Hui, X.; Zhang, W. A Novel Thermal-Fluid Topology Optimization of the Frame Mold for Composite Autoclave Process. *Appl. Therm. Eng.* **2024**, *247*, 123050. [CrossRef]
59. Upadhya, A.R.; Dayananda, G.N.; Kamalakannan, G.M.; Ramaswamy Setty, J.; Christopher Daniel, J. Autoclaves for Aerospace Applications: Issues and Challenges. *Int. J. Aerosp. Eng.* **2011**, *2011*, 1–11. [CrossRef]
60. Charchalis, A.; Kneć, M.; Żuk, D.; Abramczyk, N. Use of 3D Optical Techniques in the Analysis of the Effect of Adding Rubber Recyclate to the Matrix on Selected Strength Parameters of Epoxy–Glass Composites. *Acta Mech. Autom.* **2023**, *17*, 333–346. [CrossRef]
61. Lusiak, T.; Knec, M. Use of ARAMIS for Fatigue Process Control in the Accelerated Test for Composites. *Transp. Res. Procedia* **2018**, *35*, 250–258. [CrossRef]
62. Falkowicz, K.; Ferdynus, M.; Rozylo, P. Experimental and Numerical Analysis of Stability and Failure of Compressed Composite Plates. *Compos. Struct.* **2021**, *263*, 113657. [CrossRef]
63. Rozylo, P.; Falkowicz, K. Stability and Failure Analysis of Compressed Thin-Walled Composite Structures with Central Cut-out, Using Three Advanced Independent Damage Models. *Compos. Struct.* **2021**, *273*, 114298. [CrossRef]
64. Jonak, J.; Karpiński, R.; Wójcik, A. Numerical Analysis of Undercut Anchor Effect on Rock. *J. Phys. Conf. Ser.* **2021**, *2130*, 012011. [CrossRef]
65. Karpiński, R.; Jaworski, Ł.; Jonak, J.; Krakowski, P. Stress Distribution in the Knee Joint in Relation to Tibiofemoral Angle Using the Finite Element Method. *MATEC Web Conf.* **2019**, *252*, 07007. [CrossRef]
66. Jonak, J.; Karpiński, R.; Wójcik, A. Numerical Analysis of the Effect of Embedment Depth on the Geometry of the Cone Failure. *J. Phys. Conf. Ser.* **2021**, *2130*, 012012. [CrossRef]
67. Rogala, M.; Gajewski, J. Analysis of the Triggering Mechanism of the Square Thin-Walled Absorber. *Adv. Sci. Technol. Res. J.* **2023**, *17*, 206–216. [CrossRef]
68. Jonak, J.; Karpiński, R.; Wójcik, A.; Siegmund, M. The Effect of Undercut Anchor Diameter on the Rock Failure Cone Area in Pullout Tests. *Adv. Sci. Technol. Res. J.* **2022**, *16*, 261–270. [CrossRef] [PubMed]
69. Jonak, J.; Karpiński, R.; Siegmund, M.; Machrowska, A.; Prostański, D. Experimental Verification of Standard Recommendations for Estimating the Load-Carrying Capacity of Undercut Anchors in Rock Material. *Adv. Sci. Technol. Res. J.* **2021**, *15*, 230–244. [CrossRef] [PubMed]
70. Jonak, J.; Karpiński, R.; Wójcik, A.; Siegmund, M.; Kalita, M. Determining the Effect of Rock Strength Parameters on the Breakout Area Utilizing the New Design of the Undercut/Breakout Anchor. *Materials* **2022**, *15*, 851. [CrossRef] [PubMed]
71. Teter, A.; Kolakowski, Z. Buckling of Thin-Walled Composite Structures with Intermediate Stiffeners. *Compos. Struct.* **2005**, *69*, 421–428. [CrossRef]
72. Li, Z.; Cen, S.; Wu, C.; Shang, Y.; Li, C. High-performance Geometric Nonlinear Analysis with the Unsymmetric 4-node, 8-DOF Plane Element US-ATFQ4. *Numer. Meth Eng.* **2018**, *114*, 931–954. [CrossRef]
73. Zienkiewicz, O.C.; Taylor, R.L. *The Finite Element Method*, 5th ed.; Butterworth-Heinemann: Oxford, UK, 2000; ISBN 978-0-7506-5049-6.

74. Sun, B.; Zeng, Z.; Zhang, Y.; Shen, W.; Ou, J. An Effective Geometric Nonlinear Analysis Approach of Framed Structures with Local Material Nonlinearities Based on the Reduced-Order Newton-Raphson Method. *Soil Dyn. Earthq. Eng.* **2023**, *172*, 107991. [CrossRef]
75. Falkowicz, K. Experimental and Numerical Failure Analysis of Thin-Walled Composite Plates Using Progressive Failure Analysis. *Compos. Struct.* **2023**, *305*, 116474. [CrossRef]
76. Camanho, P.P.; Maimí, P.; Dávila, C.G. Prediction of Size Effects in Notched Laminates Using Continuum Damage Mechanics. *Compos. Sci. Technol.* **2007**, *67*, 2715–2727. [CrossRef]
77. Rozylo, P. Failure Phenomenon of Compressed Thin-Walled Composite Columns with Top-Hat Cross-Section for Three Laminate Lay-Ups. *Compos. Struct.* **2023**, *304*, 116381. [CrossRef]
78. Rozylo, P.; Debski, H. Failure Study of Compressed Thin-Walled Composite Columns with Top-Hat Cross-Section. *Thin-Walled Struct.* **2022**, *180*, 109869. [CrossRef]
79. Li, D.; York, C.B. Bounds on the Natural Frequencies of Laminated Rectangular Plates with Extension–Bending Coupling. *Compos. Struct.* **2015**, *133*, 863–870. [CrossRef]
80. York, C. Unified Approach to the Characterization of Coupled Composite Laminates: Hygro-thermally Curvature-stable Configurations. *Int. J. Struct. Integr.* **2011**, *2*, 406–436. [CrossRef]
81. York, C.B.; Lee, K.K. Test Validation of Extension-Twisting Coupled Laminates with Matched Orthotropic Stiffness. *Compos. Struct.* **2020**, *242*, 112142. [CrossRef]

MDPI
St. Alban-Anlage 66
4052 Basel
Switzerland
www.mdpi.com

Materials Editorial Office
E-mail: materials@mdpi.com
www.mdpi.com/journal/materials

www.ingramcontent.com/pod-product-compliance
Lightning Source LLC
Chambersburg PA
CBHW041950260926
51850CB00019B/2293
* 9 7 8 3 7 2 5 8 1 3 1 1 7 *